HOW MEDICINES ARE BORN

The Imperfect Science of Drugs

HOW MEDICINES ARE BORN

The Imperfect Science of Drugs

Lisa Vozza
Italian Association for Cancer Research, Italy

Maurizio D'Incalci
Istituto di Ricerche Farmacologiche Mario Negri, Italy

Translated by:
Andreas Gescher
University of Leicester, UK

NEW JERSEY • LONDON • SINGAPORE • BEIJING • SHANGHAI • HONG KONG • TAIPEI • CHENNAI • TOKYO

Published by

World Scientific Publishing Europe Ltd.
57 Shelton Street, Covent Garden, London WC2H 9HE
Head office: 5 Toh Tuck Link, Singapore 596224
USA office: 27 Warren Street, Suite 401-402, Hackensack, NJ 07601

Library of Congress Cataloging-in-Publication Data
Names: Vozza, Lisa, author. | D'Incalci, Maurizio, author. | Gescher, Andreas, translator.
Title: How medicines are born : the imperfect science of drugs /
Lisa Vozza (Italian Association for Cancer Research, Italy),
Maurizio D'Incalci (Istituto di Ricerche Farmacologiche Mario Negri, Italy)
Other titles: Come nascono le medicine : la scienza imperfetta dei farmaci. English
Description: New Jersey : World Scientific, 2017. | "Originally written in Italian"-- ECIP galley.
Identifiers: LCCN 2016055755| ISBN 9781786342973 (hc : alk. paper) |
ISBN 9781786342980 (pbk : alk. paper)
Subjects: | MESH: Drug Discovery | Technology, Pharmaceutical | Drug Discovery--history
Classification: LCC RS420 | NLM QV 745 | DDC 615.1/9--dc23
LC record available at https://lccn.loc.gov/2016055755

British Library Cataloguing-in-Publication Data
A catalogue record for this book is available from the British Library.

Desk Editors: Chandrima Maitra/Mary Simpson/Shi Ying Koe

Typeset by Stallion Press
Email: enquiries@stallionpress.com

Printed in Singapore

Foreword — Improving the Culture of Medicine

It is important to realise that improving the culture of medicine among the general public and especially patients constitutes an advantage for society at several levels. An improved understanding of health-related matters renders a patient to be more conscious of his or her rights and enables the patient to participate in decisions regarding his or her health; it enables the medical doctor to engage more confidently in dialogue and to understand the patients' needs better; and it allows the national health service to benefit from a more serene climate and to ultimately convert its plans for the future into practice in a more rational manner.

Yet, such improvement in medical culture has to be achieved in light of the fact that the citizens of today tend to extract information from the Internet, where anything can be deposited including material strongly influenced by the desire to sell new or old products. Furthermore, proper digestion of such information is difficult in the environment of an educational system which is not designed to help unravel its significance and context.

Therefore, it is very important to have documents available which have been conceived by open-minded people involved with medical matters, to improve the understanding of health-related issues. The book by Maurizio D'Incalci, Andreas Gescher and Lisa Vozza, entitled *How Medicines are Born*, has all the characteristics of a valuable component of contemporary medical culture specifically addressing drugs, which nowadays represents an important part of therapy.

The authors take the readers by the hand and guide them through the labyrinth of the procedures, problems, and especially details of the research associated with the birth of new drugs.

The book is a good and engaging read, the language is simple and avoids any pomposity. This is reflected by the subtitle *The imperfect science of drugs*, a science which often exaggerates benefits and minimises risks associated with the administration of new drugs because of commercial interests. This is a science which has not yet achieved personalisation of most therapeutic treatments, and still needs to describe results in terms of probabilities of potential success. Nevertheless, it is a science which avoids resting on its laurels and instead continually strives for improvement of knowledge and insights.

D'Incalci, Gescher and Vozza speak in an independent and balanced manner and pay utmost attention to avoid creating illusions, and instead demonstrate the limits of pharmacological treatments. Therefore here are my best wishes for the success of this booklet.

It deserves many readers, because they are to benefit from experiencing both more doubts than certainties and the desire for increased insight rather than indifference towards knowledge improvement.

Silvio Garattini
Director
IRCCS — Mario Negri Institute for Pharmacological Research
Milan, Italy

About the Authors

Maurizio D'Incalci is a medical doctor, specialised in pharmacology and oncology, who directs the Department of Oncology at the Mario Negri Institute for Pharmacological Research in Milan. He has worked for two years in the Molecular Pharmacology Laboratory, Division of Cancer Treatment of the National Cancer Institute in Bethesda, Maryland. The field of his research is the pharmacology of anticancer compounds. He has contributed to the identification and characterisation of the mechanism of action and the preclinical and clinical development of several drugs that are now used in the clinical therapy of tumours. He is the author of 500 scientific publications and several book chapters. In 2007, he was made visiting professor at the UK Open University, and he serves on the editorial boards of several international scientific journals, and is a member of the scientific committee of the Italian Association for Cancer Research (AIRC), and several other foundations and ethics committees of Italian and international scientific institutes. He has been chairman of the Research Division of the European Organisation for Research and Treatment of Cancer (EORTC) and a member of many research grant evaluation bodies instituted by Cancer Research UK and the UK Department of Health and Social Security.

Andreas Gescher (translator) is Emeritus Professor of Biochemical Toxicology at the University of Leicester, UK. A pharmacist by training, he obtained his PhD in Pharmaceutical Chemistry from the University of Frankfurt, Germany, and spent the formative years of his academic career in the Experimental Cancer Chemotherapy Research Group at Aston University in Birmingham as a member of the research team which was then engaged with the discovery and development of the anticancer drug temozolomide. Extended research stays at the Mario Negri Institute in Milan, Italy, and the Department of Biochemistry and Biophysics at Oregon State University in Corvallis broadened his academic horizon. From 1993 onwards, he was a section leader in the UK Medical Research Council Toxicology Unit, Leicester, and from 2002–2013 Codirector of the Cancer Biomarkers and Prevention Research Group in the Department of Cancer Studies at Leicester. His research interests have been the pharmacology and toxicology of cancer chemotherapeutic and cancer chemopreventive agents. He has published around 150 scientific papers and mentored ~40 graduate students. He has been chairman of the British Association of Cancer Research and the Pharmacology and Molecular Mechanisms Group of the European Organisation for Research and Treatment of Cancer.

Lisa Vozza is a well-known Italian science writer, trained in biology, and Chief Scientific Officer at the Italian Association for Cancer Research (AIRC). Her books include: "Nella mente degli altri" (*In the Mind of Others,* Zanichelli, 2007) with Giacomo Rizzolatti; "I vaccini dell'era globale" (*Vaccines in the Global Era,* Zanichelli, 2009, Galileo Literary Prize 2010) with Rino Rappuoli; "Come nascono le medicine" (*How Medicines are Born,* Zanichelli, 2014) with Maurizio D'Incalci. Besides writing books, she is the editor of a series of popular science books, Chiavi di lettura, and the author of over 170 articles on her blog, Biologia e dintorni. She is often invited to present her books in schools, bookshops and other venues, and to teach science writing as a complementary skill for scientists.

In parallel to her work as science writer, she is in charge of the peer review process at a major cancer charity in Italy: each year, thousands of cancer research proposals are evaluated by a selected group of 600 international experts, in line with the standards and practices of the most advanced research charities. In the past, she worked for the European editions of Scientific American.

Precautions and Warnings before Using this Book

The authors advise those who expect a catalogue of drugs or an encyclopaedia of medical advice to stop reading immediately. More than 3,000 new drugs have been in development in 2015 for the indication of cancer alone. Therefore, writing about each and every medicine on the market or about all diseases would be an undertaking beyond the ability and desire of these authors.

The book is absolutely contraindicated for those who are after certainties, because the read is unlikely to elicit the desired effect in such individuals. The science of drugs is highly imperfect and furnishes results, the implications of which are inevitably uncertain and preliminary rather than permanent and guaranteed.

Reading is recommended to those with a disposition to consider the notion that our knowledge of diseases and ways of treating them is rather restricted, although it is continually developing and expanding.

A potential persistent side effect of this book in those who dare venture into its pages might be the ineradicable insertion of a particular idea in the reader's mind with the danger of it ultimately infecting the whole brain. This idea entails that the development of each new drug consists of a series of tests and experimental steps, the outcomes of which are often highly uncertain and dubious.

Those who suffer from allergies against small letters may have to take appropriate precautions, as capital letters have been completely avoided for the generic and commercial names of drugs mentioned in this book.

The cover of this book lacks an expiry date because it is likely that much of the information contained in it will be outdated within a decade from now if not earlier.

If you are among the reckless individuals who are not turned off from reading this book by the above warnings, precautions and contraindications, please take up the challenge and read on. The authors hope you will enjoy the read, and they assume full responsibility for what they have put on paper, which is within the limits of their undoubtedly modest knowledge. They apologise already here for any errors and ambiguities.

Contents

Chapter 1
To Cure, Heal or Repair

The road to new drugs can follow very diverse directions. It can be straight and quick like a motorway, but more often it is arduous and tortuous, accompanied by mirages, unexpected deviations and scattered obstacles along the way, tormenting pharmacologists today as they did yesterday.

In this little booklet, we describe several routes, some easy, some littered with obstacles, which have led to medicines now available in the pharmacy, and which collectively impart some extra life or improved health on the sick and suffering. You will find out about famous medicines, from ancient aspirin to modern pharmaceuticals which control blood pressure, now used by millions of people every day, and also about some recently developed antitumour agents. *En route* there will be antivirals, antidepressives and other intriguing molecules. We want to tell you also about drugs which, although less well-known, are interesting because their development involved enlightening or instructive stories.

Together with such drugs you will meet inquisitive people, who, inspired by their desire to cure patients or their passion for chemistry, have provided us with the prodigious number of medicines which we now have, as compared to only a century ago.

Unresolved Medical Problems

Beyond specific events which may have influenced the development of this or that new drug molecule, most novel drugs share a fundamental element at their invention stage: an unresolved medical problem.

Such a problem can be a disease such as Alzheimer's, which is still most difficult or impossible to cure, or pathologies such as those associated with tumours, which can now be controlled to some extent, but by no means totally. Alternatively, an unresolved medical problem can arise from a need created by society rather than a disease, like female contraception. Such problems and necessities have guided the development of thousands of medicines which one finds today neatly stacked on the shelves of every pharmacy. Now we invite you to follow us. There is a lot to discover browsing through these shelves.

Drugs which Cure

There are about 14,000 different packages of remedies in a typical pharmacy, but only a small portion of these can eliminate the cause of a disease. The overwhelming majority treats symptoms.

Antibiotics are among the pharmaceuticals known for their ability to cure a disease at its roots, given it is caused by a bacterium. You may perhaps have yourself taken an antibiotic the last time your doctor prescribed one for you to treat a respiratory infection which provoked greenish mucus, fever or a very bad cough. Antibiotics do eradicate more or less all bacteria which can cause infections, although they are unable to kill viruses.

We have known antibiotics for about a century. To be precise, since 1928, when Sir Alexander Fleming observed by chance that the growth of bacteria on a Petri dish contaminated with a mould was inhibited. Interestingly, the notion that moulds produce antibacterial substances had already been proffered earlier in Naples, more than 30 years before.

In 1890, Vincenzo Tiberio, a medical officer in the Italian Navy, observed that some neighbours of his suffered from terrible diarrhoea after green moulds had been removed from the wall of a spring from which water was taken. Tiberio went on to publish, in a contemporary periodical, the hypothesis that some antibacterial substances secreted by the mould covering the well may have protected the intestinal health of his neighbours: the removal of this mould may have rendered them susceptible to the deleterious action of bacteria contained in the water. Nobody, however, showed any interest to follow his intuition up, nor did Fleming's observation cause any stir, at least initially (Fig. 1).

It was the large number of soldiers wounded by firearms during the Second World War, with the overwhelming need to stop infections they contracted, which provoked the rediscovery of penicillin and its industrial production, in addition to the discovery of other antibiotics. It is therefore not coincidental that Fleming should have received the Nobel Prize in 1945, a good 18 years after the initial discovery of penicillin, when the world wide conflict was about to draw to a close. Although penicillin had only just about entered general practice in 1944, it saved as many as millions of human lives even during those early months of use.

Figure 1. Alexander Fleming (left, Wikipedia) and Vincenzo Tiberio (right, Wikipedia).

Between then and now, the number of people who have survived an infection thanks to antibiotics is incalculable. It helps imagine what the world was like prior to these drugs: Five women in a thousand died giving birth, one person in nine caught a severe skin infection from a harmless scratch or insect bite, and three people in ten fell ill with pneumonia. Most of those who did not die from these ailments failed to recover fully. After an ear infection, a patient tended to remain deaf, and after a throat infection, one often developed heart disease.

Try to go back in time to the preantibiotics era and ask yourself: Would you want to have a wisdom tooth pulled out or an appendix removed? Not to mention open-heart surgery, organ transplantation or intensive therapy? All of these would be unimaginable manipulations.

Even in our days, this horror scenario is no longer so far-fetched because many common bacteria no longer respond to antibiotics. In order to understand how such insensitivity arises, one can conduct a very simple experiment. It entails growing a population of bacteria in culture and exposing them to an antibiotic. Almost all the bacteria will die except a small group of resistant cells. Now what about these survivors? With respect to the large number of dead bacteria, the surviving ones have a genetic mutation which renders them capable of withstanding the onslaught of the antibiotic. The mutation itself would only be active in a single bacterium and its progeny, were it not for a further cunning ability of the microbes. They can build a little temporary biochemical bridge which allows two neighbouring bacteria to communicate and pass a piece of mutated genetic material from one to the other. Resistance is thus transmitted not only vertically, to "daughter" bacteria, but also horizontally, to "distant relatives" and "friends".

The abundant use of antibiotics not only in humans, but also in farm animals, has led to the selection of many strains of bacteria which are able to get round the activity of antibiotics. Due to this advantage, these strains reproduce themselves more amply than their antibiotic-sensitive "cousins", leaving medics and the sick with drugs turned blunt weapons. Nowadays, antibiotics need to be taken at higher doses and for longer periods of time as compared to the past, and relapses are more frequent. Resistant bacteria are robust and constitute living proof for the extraordinary capacity of biological species to adapt to the most disadvantageous situations.

New drugs against bacteria are urgently needed. Given current technologies, it should indeed be possible to discover them, but few are emerging on the horizon. For commercial reasons research into antibiotics is low-key. An antibiotic is cheap, cures rapidly and effectively (often one package suffices), and it becomes obsolete within a short time, because resistant strains emerge rapidly. In the light of the high costs of drug development, the pharmaceutical industry tends to aim for new drugs against chronic conditions which patients have to take for life (meaning many packages of drugs are required!). This argument is short-sighted, because in a world without antibiotics, the chronically ill would also be at risk. But as it stands, some companies seem to want to make their money in this fashion.

But there is good news. In 2010, the US government offered the pharmaceutical industry financial incentives to stimulate the development of novel antibiotics. And since then similar examples followed in European countries. These proposals are a sure sign that the concern with the danger of resistant bacteria to public health is high. Let us hope that we can see the fruits of these initiatives soon.

Drugs which Can Heal

Different from the antibiotics, the great majority of drugs cures symptoms, but does not remove the cause of the disease, such that this does not clears up the disease completely. That said, the result of at least two centuries of pharmacological research and several thousand years of medical research seem modest, as the pharmacologist's and medic's aspirations are undoubtedly to eliminate diseases *together* with their causes. Nevertheless, let us not underestimate how important it can be to evade even only the symptoms of illnesses and the complications which they provoke.

Think of insulin, a protein which is usually produced in cells of the pancreas, and which controls the concentration of sugar when released into the blood. Those who generate insufficient quantities of insulin fall ill with diabetes, a fatal disease up to less than a century ago, and a common one — we all know at least one person with diabetes.

From Animal Insulin to the Biotechnologically Generated Hormone

The situation for people with diabetes was dramatically improved at the beginning of the last century, due to the tenacity of a Canadian medic by the name of Frederick Banting. Like many medics and scientists at that time, Banting wanted to isolate insulin. But the protein turned out to be difficult to extract from the cells of Langerhans in the pancreas, in which it is produced, because it is rapidly degraded by trypsin, an enzyme made by neighbouring cells.

Banting acquired the knowledge of a procedure which avoids this problem. It turned out to be sufficient to block the so-called pancreatic duct, a channel in the pancreas, so that once this duct was tied up, the passage of trypsin would stop, thus leaving the Langerhans cells intact.

In order to do the experiment and test the procedure, Banting needed a laboratory and an assistant. Taking advantage of the summer (even laboratories are sometimes empty!) he asked a Scottish biochemist and physiologist, John Macleod, who worked in Toronto, for help. Macleod offered him not only the laboratory, but also a generous endowment in the form of an assistant, who had to be chosen from among two youngsters in the lab, Charles Best and Clark Noble.

In addition he provided ten dogs with which to conduct the experiments. Best and Noble tossed a coin to determine who would gamble away the vacation. It was Best's turn to stay in the lab, and he helped Banting to isolate insulin from the dogs' pancreas.

About 422 million patients with diabetes world-wide who can survive today, their families, their friends — and in fact all humans — must feel immense gratitude, because there hardly exists a family untouched by diabetes. Gratitude towards Frederick Banting for his tenacity and to Charles Best for his lost holidays, who justly received half of the money which came with the Nobel Prize from the most conscientious and generous Banting. Gratitude as well to the ten very precious dogs, unwitting but essential allies in the fight against this unforgiving illness (Fig. 2).

Today, we no longer have to sacrifice animals to offer insulin and longevity to patients with diabetes. This fact is owed to the English biochemist Frederick Sanger, who was the first to identify

Figure 2. Charles Best and Frederick Banting with one of the dogs whose sacrifice helped isolate insulin (University of Toronto, School of Medicine).

the structure of insulin, thus pioneering the road to the first synthetically produced hormone.

The insulin which diabetics receive today is another miracle of research. It is no longer produced in a chemical industrial plant but instead in a biological microfactory, or rather a bacterium or yeast, into which the human insulin gene has been inserted. Credit for this achievement belongs to scientists from Genentech, one of the first biotechnology companies which was founded in 1976. The product is called "recombinant" insulin, because it is the product of the genetic recombination between genes from humans and microorganisms.

A short aside here: If you suffered from diabetes, would you refute recombinant insulin because it is a product of a genetically modified organism (GMO)? The procedure to make a microorganism produce insulin is the same as that necessary to create a GMO plant, involving insertion of genetic material from one species into another. Issues like this need to be considered by the opponents of foodstuffs generated by GMO.

"Anything but Chemistry!"

Apart from insulin, there are many other pharmaceuticals, which by curing only symptoms can prevent complications and reduce mortality in a drastic fashion. An example are the antihypertensive drugs, remedies which control arterial pressure and are prescribed for people in whom maximal and minimal blood pressure values are above 140/90 mmHg, respectively. It is thought that in the so-called developed world about one in three people suffer from hypertension. Since the introduction of antihypertensives into therapy, the mortality of stroke, infarction and other heart problems has been reduced by as much as 30–40%.

Among the drugs most frequently used in the treatment of hypertension are the so-called angiotensin converting enzyme (ACE) inhibitors, molecules which interfere with hormonal regulators of body fluid volume and thus of blood pressure. Angiotensin is one of the hormones, the activity of which is affected by the drugs.

The inventor of these drugs is Sir John Vane, who won the Nobel Prize for Physiology or Medicine in 1982 for yet another discovery, perhaps an even more important one than his first — the understanding of the mechanism of action of aspirin, one of the most frequently used drugs in the world.

Before we focus on aspirin, let us embark on a slight diversion concerning Sir John and his brilliant pharmacologist colleagues. It is easy to think that these scientists would be super-human, fundamentally different from us mere mortals. In reality, the greatest scientists are guys riddled with worries and experiencing moments of severe discouragement and serious doubt. In his speech during the Nobel Prize award ceremony, John Vane told an instructive story. After his degree in chemistry at Birmingham University he was asked what he intended to do afterwards, to which he responded "anything but chemistry". So even the best are sometimes disheartened! The important thing is to pick oneself up again. Let us now return to aspirin.

The Inheritance of the Willow Tree

Acetylsalicylic acid, commercial name aspirin, is an effective drug against fever, pain and inflammation. It has been used for at least a century, or rather since the German chemist Felix Hoffmann, an employee of the chemical company Bayer AG, came up in 1897 with this more tolerable version of the equally efficacious but somewhat toxic salicylic acid extracted from the bark of the white willow tree.

Already the Sumerians and the ancient Egyptians recognised the capacity of certain constituents of this plant to combat fever and to alleviate pain. Among the thousands of substances to which the curative potency of the willow could be attributed, are the salicylates. They are among the very few ancient remedies which have successfully passed modern scientific examination and of which we now know the mechanism of action.

In 1980, John Vane discovered that acetylsalicylic acid inactivated some enzymes called cyclooxygenases (COX) in an irreversible mode. COX enzymes enable the production of prostaglandins and thromboxanes, two families of molecules involved in inflammation and haemostasis, respectively. Haemostasis is the process which stops bleeding caused by a wound through the formation of a kind of plug made of blood cell fragments. These are called platelets and they form an aggregate also requiring a protein called fibrin.

Eighty years passed from the synthesis of acetylsalicylic acid to the experiments of Vane, encompassing more or less a century of medical prescriptions against fever and inflammation without anybody knowing how aspirin worked.

You probably know what fever is: an increase in body temperature above the normal value of 36.5–37.5 °C, one of the most common manifestations of an underlying medical problem. Inflammation, which occurs frequently, is the response of the organism to dangerous stimuli like an irritant substance, an infection caused by microorganisms or the presence of damaging cells. Inflammation reduces the problem and initiates the healing process. The classic symptoms of inflammation are pain, heat, skin reddening and swelling, symptoms which sometimes become severe and need to be blocked. Now, aspirin works like this: By effectively inhibiting the COX enzymes it counteracts the production of prostaglandins, which would augment the inflammation and make the situation worse. Other drugs often employed against inflammation, such as ibuprofen, also inhibit COXs. Together with aspirin they belong to the group of non-steroidal antiinflammatory drugs (NSAID), although the inhibition they cause is reversible, a difference which renders aspirin unique.

From Aspirin to Baby Aspirin

Ask anyone older than 50 whether he or she takes aspirin every day. At least some of them will tell you that they do. If you investigate a bit further, you might find out that it is a special aspirin they take, low dose acetylsalicylic acid, known as "baby aspirin" or "aspirin cardio".

It has been used in every-day medicine for about the last twenty years, and very many elderly people are being prescribed baby aspirin in order to prevent so-called thrombi, solid clots made up of red and white blood cells, platelets and fibrin, which block the circulation and cause permanent damage when formed in the blood vessels, heart or brain. According to a recent paper in the *American Journal of Preventive Medicine*, 52% of US citizens aged between 45 and 75 years take aspirin regularly with the intent to prevent disease, predominantly infarction and stroke. And a further 21% had used it at some point in the past. So it is not surprising that the total annual consumption of aspirin world-wide is around 100 billion tablets.

The development of a drug to prevent blood clots in the arteries was initiated at the end of the 1970s, when it was found that thromboxane, a substance produced by the platelets, favours blood clotting, whilst prostacyclin, generated by endothelial cells of the blood vessels, opposes it.

Around 1982, two 30-year-old pharmacologists, Carlo Patrono at the Catholic University of Rome and Garret FitzGerald at Vanderbilt University in Nashville, Tennessee, established in their laboratories at the same time that low-dose aspirin (75–100 mg) could inhibit thromboxane synthesis selectively. The results of subsequent clinical trials rendered a novel therapy to prevent blot clotting possible and also changed the treatment strategy for patients with heart conditions such as infarction or stroke. Patrono and FitzGerald received many prizes for their discovery, the most recent in 2013 being the Lefoulon-Delalande prize, considered the most prestigious in the world in the area of cardiovascular research.

Drugs which Break Up Preformed Blood Clots

A heart attack or infarct is the death of heart tissue which potentially kills the patient, consequent to a blood clot occluding an artery through which blood carries oxygen and nutrients to the heart.

Is it possible to reduce the risk of death by pharmaceutical intervention after a heart attack? In order to answer this question, a group of Italian researchers conducted a clinical trial from 1984–1985 in 12,000 patients recruited from the coronary units of

172 hospitals. Streptokinase, a fibrinolytic drug, was chosen for study because it was known at the time to be able to dissolve blood clots. What was not known, however, was whether it would actually work in patients who had suffered a heart attack. The trial demonstrated that streptokinase injected into patients within 12 hours after the first symptoms of a heart attack did indeed reduce mortality by 20%. "...The secret of the success of the study was not only the innovative ideas of the researchers, but also the rigour with which the physicians conducted the trial recruiting a very large number of cases..." commented a cardiology expert from Oxford University in 2004 in an editorial of the *American Heart Journal*. Today antifibrinolytic agents have been used for years in hospitals around the world and have saved many thousands of lives.

When Serendipity Leads to New Drugs

Some great discoveries are not the result of a brilliant idea, but of surprising and unexpected observations unrelated to the research objectives. This is referred to as serendipity, a term coined in 1754 by Horace Walpole, an English politician and man of letters who thus defined the ability to make fortunate discoveries by chance. The term is still used today in science. It stems from an ancient Persian fable in which three princes of the kingdom of Serendip (the Arabic name for Sri Lanka) "were always making discoveries, by accidents and sagacity, of things they were not in quest of", as described by Walpole.

In 1965 Barnett Rosenberg, an American chemist at Michigan State University, attempted to find out if an electric current could affect the proliferation of bacteria. To that end, he designed a flask in which a current flowed between two platinum electrodes dipped in a solution containing growing bacteria. Rosenberg noticed that once the current was switched on, the bacteria stopped multiplying and dividing. The block of cell division thus discovered was not due to the current as such, given the fact that the effect persisted even in the absence of the current. Instead the cause was the reaction of platinum with a salt in the solution. They formed a compound, called cis-platinum, which was able to damage the DNA of the bacteria and thus interfere with their normal growth.

Subsequently Rosenberg isolated the agents present in the solution among which only cis-platinum, apart from blocking bacterial growth, was capable of stopping the proliferation of certain tumour cells.

Was Rosenberg simply lucky? No doubt luck helped him. But his discovery was also the result of intelligent observation linked to robust chemical training which enabled him to identify the substance responsible for his unexpected observation.

Important serendipitous discoveries can also be made in the clinic. For example, viagra was used experimentally in patients with a heart condition, because it had been shown that it could act upon the vascular smooth musculature and affect blood fluidity. The idea was that it might prevent the closure of coronary arteries, thus averting a heart attack. The experiments did not deliver the expected results with respect to cardiac disease, but some male patients noted that the treatment provoked an unexpected side effect, a strong tendency towards penile erection.

Drugs Discovered to a Plan

There are drugs which have been discovered by chance, and their mode of action has been discovered only after a century of use. Then there are drugs which have been designed right from the start to function in a particular way. One such drug is imatinib, the commercial name of which is gleevec. It is arguably one of the most famous anticancer drugs in use. It cures only symptoms but has changed the history of chronic myeloid leukaemia (CML), an illness in which a population of white blood cells grows out of control.

The story which led to imatinib started in 1973. It is evening and we are in the house of Janet Rowley, a 47-year-old geneticist and medical doctor who works part-time at the University of Chicago whilst raising four sons full-time. In the laboratory, Dr Rowley is a specialist in chromosome staining. This technique, which she learnt in 1970 during a sabbatical year at Oxford University, allows delineation by different colours of the segments of each of the 23 copies of the stick-like macromolecular structures which contain the genetic material inside the nucleus of each human cell. Given the degree of resolution which prior to the introduction of this technique was unthinkable, the staining permits precise differentiation between the chromosomal segments in different cells, for example, healthy as compared to diseased ones.

Sitting at the table of her kitchen after having cooked a meal, cleared the table and taken the children to bed, Dr Rowley looks at some photographs after microscopic enlargement of stained chromosomes which belong to white blood cells from patients with CML. Among the 23 chromosomes of these cells, there was an anomaly which was repeated in each patient. Chromosome 22 was always shorter than normal, which had already been observed in the 1950s by two pathologists from Philadelphia, who had given the anomaly the name of their city (Philadelphia chromosome). The problem was that nobody had succeeded to understand where the broken-off head of this chromosome had ended up. That is, nobody except Dr Rowley, who that evening, sitting at the table of her kitchen looking at these photographs of stained chromosomes, observed that the Philadelphia chromosome was the aberrant result of a fusion (a "translocation" to use the geneticists' jargon) between the tail of chromosome 22 and the head of chromosome 9.

This was the first time that a chromosomal anomaly was suggested to be the cause of a tumour. Dr Rowley's discovery provoked initially a wave of scepticism, then for a period what one may call "puzzled tolerance", and ultimately a revolution. Now we know that several cancers, not only CML, can be caused by translocations of genetic material from one chromosome to another. Janet Rowley (Fig. 3) received more or less every recognition available for her discovery except the Nobel Prize. Why she would not have received the phone call from Stockholm remains a mystery.

But let us return to the fusion between chromosomes. How can it cause CML, and which are the genes involved? The hunt for the mechanism was on. In 1982, a group of Dutch scientists isolated a

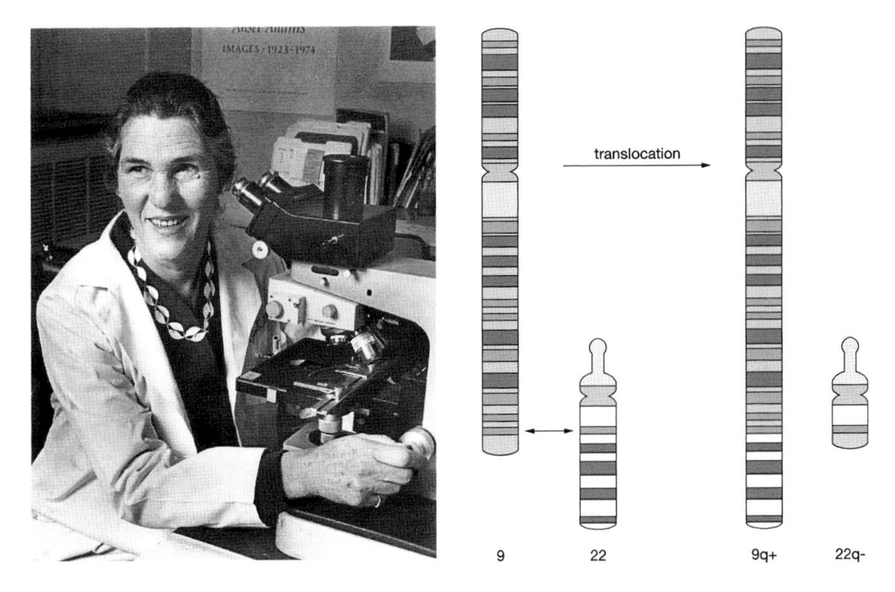

Figure 3. Janet Rowley (left, with the kind permission of University of Chicago Maroon) and the translocation between chromosomes 9 and 22 (right, redrawn from: National Library of Medicine), creating the Philadelphia chromosome, the cause of chronic myeloid leukaemia.

gene on chromosome 9 and named it *Abl*, whilst another group in Maryland in the United States found its fusion partner on chromosome 22, and it was named *Bcr*. But the proof of the pudding was provided in 1987 in the MIT laboratory of David Baltimore, an American biologist and Nobel Prize winner. The *Bcr-Abl* gene was inserted into the white blood cells of a mouse, and shortly after the infusion of the thus "engineered" cells, the mouse came down with a leukaemia very similar to human CML.

Now it remained to be explained how this fused gene can cause CML. In Baltimore's laboratory the product of the *Bcr-Abl* gene was found to be a kinase, that is a kind of switch capable of activating or inactivating other proteins by attachment of phosphate groups. But *Bcr-Abl* is not an ordinary kinase, rather an exuberant protein which by activating other proteins forces cells to continue dividing.

Finally, CML is no longer a mystery. To recapitulate: the fusion between two particular genes in white blood cells creates an aberrant gene, which in turn generates a hyperactive protein switch. The result is the uncontrolled proliferation of a portion of the white blood cells which overwhelm all the other blood cells and cause the leukaemia.

To unravel a biological mechanism engaged by a tumour provides a tremendous intellectual satisfaction, but in order to cure the disease drugs are required. Thus pharmacologists got to work as soon as the identity of the *Bcr-Abl* kinase and in particular its molecular shape had been clarified.

Fortunately, the *Bcr-Abl* molecule contains a cavity into which a small molecule can be inserted to block protein function. Nick Lydon and Alex Matter, two biochemists at Ciba-Geigy in Basel, a pharmaceutical company of that time, tried to construct a molecule which could inhibit *Bcr-Abl* in a selective fashion among the hundreds of useful or innocuous kinases present in the organism.

The experimental attempts generated a series of inhibitory drug candidates, which needed to be tested in patients in order to find out whether they can block the disease without toxicity. So at the end of the 1980s, Lydon went across the ocean to the Dana–Faber Cancer Institute in Boston, one of the most advanced cancer hospitals in the United States with a great tradition of leukaemia treatment research. Here Lydon met Brian Druker (Fig. 4), a young haematologist engaged with the treatment of CML, up to that point an incurable disease. Druker proposed a clinical experiment to the Dana–Farber Institute to test in collaboration with Ciba-Geigy the molecules developed by Matter and Lydon. However, for legal reasons the collaboration did not take off. Thus in 1993, the frustrated Druker left the prestigious Bostonian institute and the Atlantic coast and moved to a humdrum clinical centre in Oregon at the Pacific coast.

In the meantime, the majority of CML patients continued to die, the only treatment available being a very complicated bone marrow transplant, which caused severe side effects and could only be offered to fit patients for whom a compatible bone marrow donor was available.

But Druker did not give up in Oregon either, as he was hell-bent on improving his patients' lot. Resuming his collaboration

Figure 4. Brian J. Druker, the father of imatinib, the first drug to block a specific cancer target (with the kind permission of Brian Druker).

with Ciba-Geigy, he started to experiment on one of the molecules which he had obtained from Switzerland, first in cells and then in mice with CML. The results were astonishing. Both in cells and in mice, all tumour cells died within a few hours of treatment. Druker and Lydon published their results in the prestigious scientific journal *Nature Medicine*. Within a short time CML patients should have a drug, or so Dr Druker thought.

But the road to discovery turned out to be an even longer one for him and his patients. Ciba-Geigy merged with Sandoz, another pharmaceutical company, to create Novartis, a huge company still around. In 1993, the industrial giant was undecided as to whether it should invest several millions of Swiss francs to investigate this drug with the expectation of only a low economic return. As things stood, it would only be used to treat patients with CML, who make up just 10% of all forms of leukaemia, slightly more than 5,000 patients per year. But Druker did not give up, and after five long years he succeeded in 1998 to convince Novartis to produce several grams of this molecule which enabled him to conduct a small preliminary study in 100 patients. Half of these received the drug.

The results were breath taking: In 53 of the 54 patients who received the treatment, the leukaemia seemed to have vanished after treatment for only several days. In reality tumour cells did persist, but they were no longer capable of reproducing themselves at the frantic speed which had made them so dangerous. The drug rendered the leukaemia cells actually innocuous.

Since then this drug, named imatinib or gleevec, has become standard therapy for CML and other leukaemias and lymphomas in which the Philadelphia chromosome has been identified.

Imatinib blocks the lethal consequences of a problem, the fusion of two chromosomes, but cannot restore the two chromosomes to their prefusion state. Unfortunately, nobody has yet found out how to correct genetic damage *per se* using a pill.

So imatinib is a drug which blocks a defective mechanism, but does not repair the problem causing it. Therefore the drug needs to be taken for life with the concurrent risk of potential side effects on the heart and of developing other tumours.

Yet, the alternative is death. The compromise — taking the drug and accepting the risk — is more than acceptable to the over 70,000 patients in the world who today survive with CML. Thanks to imatinib, leukaemia has been transformed from a lethal to a chronic disease.

Imatinib is the first antitumour drug designed in a rational and specific fashion on the basis of the knowledge of the biological mechanism which drives the disease. For the last 15 years, at least, pharmacologists have attempted to duplicate the success of gleevec in other tumour types. But this is not easy, because in most cancers unique and relatively simple anomalies like the Philadelphia chromosome in CML, are the exception rather than the rule.

Finally, also the miraculous imatinib has limitations. For example, it works in about nine out of ten patients. Moreover, many patients in whom imatinib is effective develop resistance after some time. You may ask what is different about the unfortunate patients who fail to respond?

Their leukaemia cells are likely to possess certain biological characteristics which render them insensitive towards imatinib, irrespective of the fact that the disease appears to be the very same as in those patients who respond to it. It is also conceivable that the cells of these patients harbour mutations which modify the cavity of the *Bcr-Abl* protein, into which imatinib can insert, just like when a lock is changed, the old key no longer works. The reality is that even for one of the overall best understood diseases like CML, there are a lot of things which we do not yet understand. We do not know why certain patients fail to respond to imatinib, and even before having administered the drug, we cannot predict who will respond and who will not. Each patient constitutes a unique combination of genetic and molecular characteristics which are never completely super-imposable on another patient, even his or her twin sibling.

Towards Drugs which Can Repair?

There are many ways in which medicines can be classified. In this chapter, we have described a few major categories: medicines which cure, those which heal symptoms, medicines discovered by chance, and those the design of which was planned as part of the search for a cure of a disease with a well-understood mechanism.

The histories of most successful new medicines resemble that of aspirin rather than that of gleevec still today. Drugs have been chosen from the many molecules present in nature and reproduced or modified in a medicinal chemistry laboratory. Then they have been tested against various diseases. When found to work and to lack unacceptable toxicity, they have been introduced into clinical practice. Only in rare cases has the fundamental understanding been gained as to why they are efficacious and how they exert their pharmacological activities.

Today pharmacologists try to develop molecules in a fashion resembling the discovery of imatinib, to come up with drugs designed to interfere with well-understood and molecularly defined problems. The pharmacologist's minimal objective is to try to block the damage caused by an underlying disease. But every researcher's dream is undoubtedly to repair the problem at the root of a disease, as antibiotics do so uniquely, contrasting drastically with the majority of medicines which deal only with the symptoms. Another dream, perhaps more likely one of medics rather than of pharmacologists, is to be able to identify, before treatment commences, the patients who might respond to a certain drug. This ability would avoid damaging side effects in those who are unlikely to derive any benefit from treatment.

The proverbial mountains to be climbed in order to come up with new drugs are very high, and the paths up to the top are exceedingly steep and cumbersome. Although many such mountains may ultimately be unconquerable, some summits are reachable, even though only by the most persistent climbers. Follow us, we shall attempt to show you not necessarily how one reaches the summit, but how one can at least attempt to do so, following the path from the idea to the medicine vial.

Chapter 2
Hunting for Drugs

The most up-to-date and elegant method to deliver a new medicine starts, as we have seen, with the knowledge of the mechanisms of the disease to be targeted. Such knowledge helps to design — literally at the writing desk — a drug capable of interfering with the molecules or the biological circuits which are dysfunctional or out of control.

A slight clarification here to avoid misunderstandings: The term "circuit", which you will come across several times in this book, is used in a loose and non-technical sense. It describes how biochemical, genetic, metabolic or nervous pathways, which when disorganised or hyperactive, can cause illness. Such pathways are therefore ideal targets for a drug. Back now to the rational design of drugs.

This approach is intellectually pleasing because it is linear and therefore favoured by the human brain when it comes across as a simple, logical and well thought-through strategy. Nevertheless, in science, the optimal strategy is that which actually works. The rational drug design approach has hitherto generated few successes when compared to the majority of medicines which have been discovered during the past century by the use of what one could call the "empirical method". It selects from among many chemicals the most efficacious and least toxic ones, efficacious in terms of their ability to tackle a particular medical problem. Here, questions as to their mechanisms of action are less important. Also, in the past the scientific tools to dissect disease mechanisms have been mostly lacking.

The apparent lack of success of the "rational approach" tells us two things: firstly that reality, especially in biology, is always more complicated — but also more interesting — than our imagination, and secondly that we are currently experiencing the dawn of precision medicine, which means we should be patient and wait and see how it will develop.

In this chapter, we shall find out how the hunt for new drugs is a confusing and complex adventure riddled with obstacles. However, when somebody succeeds at solving the riddle and advances a molecule to at least the stage of early clinical studies, the satisfaction is immense and the hope to be able to help desperate patients just as great. Now let us stop deviating and rather go on in a logical fashion, as the road is long, and we do not want to lose you on the way. The first step towards the development of a new drug is to know at least in part the dysfunctional circuit which causes the disease. But how many of these circuits do we know and with which degree of accuracy?

The Human Body — Is there an Instruction Manual?

Let us begin on a positive note. Whereas, until the 1970s we only knew a little of the molecular causes of diseases, nowadays, we have accumulated a veritable plethora of knowledge. The enthusiasm generated by this gigantic collection of data is considerable and the satisfaction justified, but at times we have to be mindful not to get carried away. There are, for example, some who maintain that our body is a very complex machine for which we have medicines and spare parts ready to correct each defect.

In order to understand to which extent this statement is true, a comparison with one of the most sophisticated machines which have been constructed by us humans may help — the robotic rover Curiosity. This fantastic machine, which since 2012 has been exploring the surface of Mars, knows how to do things we have hitherto not even be able to dream of, such as to reach the red planet and investigate it with the help of laboratory robotics, video cameras and other weird contraptions.

However, there are some fundamental differences between a machine like Curiosity and a human being. First of all, a robot is born with an instruction manual, because it was designed and assembled bit-by-bit by somebody. In contrast, we humans have been made in a different fashion, just like viruses, bacteria, plants and other animals. And this is the reason why the around 30,000 billion cells that make up our human "machine", were born without a plan, but emerged in a gradual manner by trial and error during the about 4 billion years of evolution of life on earth. And as nobody has constructed and assembled the cells as it has been done with the nuts and bolts, wheels and video cameras of Curiosity, we do not have an instruction manual available to consult when something in these cells goes wrong.

This is the fundamental reason why we have only a vague knowledge of the human body, the function of which remains rather mysterious. In spite of everything we have learnt especially during recent years, our "machine" is way more refined, complicated and efficient than the machines humans have been able to conceive and build. Think only of the heart, a little engine which beats about 60–70 times each minute of each hour of each month of each year for many decades, and for even more than a century in the 316,000 or so centenarians living in the world. This is an extraordinary performance, which has never been achieved in any machine designed by humans, and which is reflected in miniature in the cells which form each of our organs including the heart.

The Cell — A Great Chemical Laboratory

In a volume a thousand times smaller than a grain of rice, each cell, even the simplest one, manages to transform, assemble and take apart a myriad of elements of structures and disparate functions and work in extremely small dimensions.

A comparison between cell and human-designed machine is unfair. Because in even the most advanced laboratories, we have only been able to learn how to imitate in a crude and imprecise way some of each cell's capabilities as a natural micro workshop.

Each day a cell produces millions of molecules, mainly proteins, on the basis of the information imprinted in the DNA and conserved in the nucleus. The unit which contains each of these pieces of information is called the gene. We know about 24,000 genes, and we have learnt to read the sequence of each of these characterised by different combinations of four chemical letters called nitrogenous bases.

The genes occupy only 2% of our DNA, and the residual 98% has until recently been considered "junk" DNA — sometimes scientists tend to be contemptuous about what they fail to understand. Today, we know that what was considered junk contains pieces of DNA with very important functions for the regulation of the life of the cell. But the exploration of this little known area of bioscience has just begun; and who knows, it may furnish some surprises in the future.

Proteins are the real workers of the cell: some construct and reinforce the cell's infrastructure like builders on a construction site, others make the required energy available, others again turn switches on or off which allow or disallow encounters, interactions or reactions to occur. Then there are those which are busy with repair processes, transport, cell defence and communication. And we have not talked about those engaged in cell reproduction.

Our lack of knowledge of proteins is greater than that of the genetic material. We do know that in our body we have at least 300,000 of these extraordinary jack-of-all-trades of cellular life, although the three-dimensional shape is known for only ~6,000 of them. The knowledge of the shape of proteins, not only of the sequence of their constituting amino acid building blocks, is essential for the design of drugs targeted against them. Success in this endeavour is extremely difficult because each protein needs to form a crystal before one can study its three-dimensional structure using X rays.

Therefore many of the researchers who have succeeded in this, after having spent maybe several decades in a tough encounter with some obstreperous protein, have been rewarded with the Nobel Prize.

The Most Common Target of Drugs

By way of simplification, one might say that a drug can bind to a protein like a key fits into a lock. If the key is the drug, the lock would be the protein receptor capable of being activated to elicit a pharmacological response only if the drug is specific.

There are keys which enter a lock and allow bolting and unbolting perfectly, whilst others instead enter a lock but then get stuck and block it. In analogy, some drugs bind to a receptor and set a response in train whilst others instead only block and thus inactivate the receptor. The former are called *agonists* and the latter *antagonists*. Agonists compete with endogenous molecules like hormones binding to receptors, whilst antagonists compete with agonists restricting their ability to bind to a receptor and to cause a response.

As it happens, cells are full of receptors and molecules which can bind to them. The abundance of receptor–molecule interactions offers many opportunities for the design of drugs with specificity for a well-defined receptor. But the abundant similarities among receptors and molecules present in nature also offer many opportunities for ambiguities, that is non-specific bonds and consequences which are a principal source of unwanted side effects.

Catalogues of Genes and Proteins

Up to the 1990s, scientists worked on single genes or proteins at a terribly frustrating pace. They would have done well if they succeeded within six months or a year to unravel the information contained in a gene. To understand what this gene actually does took even more time. Let alone the resolution of the structure of a protein, for which — as we have mentioned before — even today a decade may be insufficient. You will understand that knowledge accumulated at this sluggish pace was frighteningly meagre, limited and fragmentary.

Then suddenly almost all changed: The techniques which allow identification of genes became more powerful and — particularly important — were automated permitting rapid compilation of gigantic catalogues of biological molecules. The first and most famous of these compilations has been the Human Genome Project, which within slightly less than 10 years achieved the indexing of all genes present in our cells.

This endeavour has involved collaboration between scientists from all parts of the world. It was followed by even more ambitious "omics" projects, such as inventories of all proteins which we are made up of (the so-called proteome), of all molecules involved in metabolism (the metabolome) and of all body fats (the lipidome) and so on.

Most useful to pharmacologists are those omics projects, which map out differences between a diseased and a healthy organism. The Cancer Genome Atlas, for example, is a project which from 2005 to 2014 collected all the genetic alterations contained in almost 200 diverse types of cancers, analysing the DNA from hundreds of thousands of patients from all over the world. Another example is the BRAIN Initiative ("Brain Research through Advancing Innovative Neurotechnologies") in the US, which has the objective to map out the activity of each of the 100 billion neurons operating in the human brain. The goal of this initiative is to understand what happens in the brain of people afflicted with Alzheimer, autism, schizophrenia and other cerebral problems, for which there are still few solutions.

Scientists tend not to be short of imagination or immune to fashions. So not surprisingly they have invented all kinds of omics projects in recent years, encyclopaedic efforts which attempt to list each biological variation thus satisfying the predilection for comprehensiveness dear to the collecting trait in human nature.

Although big omics projects have generated enthusiasm and great expectations, one has to be honest and say that a catalogue of molecules as such does not say anything about the function of the items indexed in it.

Furthermore, not all omics projects have really clear objectives and some lack suitably effective techniques. Therefore, some scientists ask probing questions as to the validity of these gigantic efforts.

Is it really useful to compile unending lists of genes, proteins or other molecules in order to improve our understanding of what might go wrong in our body and to find remedies? And then, among the thousands of molecules contained in these lists, how does one distinguish those capable of driving corrupted circuits from those which are there by chance as innocent bystanders?

Towards the Molecular Causes of Diseases

In order to answer these questions, which we try to do shortly, we have to turn to the diagnosis, or rather the crucial moment in which a doctor understands what the fault may be which needs to be cured. To make a correct diagnosis, that is to understand what the problem is, where it is localised and why it occurs, is the most important step towards curing a disease.

Up to the first half of the last century, diagnoses were based on what was apparent to the doctor's eye, sometimes helped by enlargement with a magnifying glass.

Within a century, we have learnt to make diagnoses in a much more precise fashion. Small quantities or tissue fragments were taken from patients for investigation in the laboratory.

Initially the pathologists, who are the doctors who make diagnoses by investigating such samples, were restricted to observe samples with minimal enlargement under the microscope. In the course of time they became aware of the fact that our tissues do not have infinite modes of appearances, and that some or other common feature can help group diseases with different causes together into a disease family.

Pathologists started to conduct experiments on small quantities of blood or tissue, using reagents which generate a certain colouration depending on the presence of a particular protein, fat or sugar in the blood or tissue. In performing these analyses pathologists have become increasingly bold, so that today they are even able to recognise specific molecules present in paltry quantities in say 10 or 20 cells.

Laboratory techniques which have become more and more important are those allowing biological imaging, like computer tomography (CTI) and magnetic resonance spectroscopy (MRI), which have generated astonishingly detailed snapshots of the inside of our body when compared to the radiographs of old.

The diagnostic power which pathologists and radiologists have managed to achieve owes a lot to the increasing repertory of omics knowledge which we have already discussed above. So to provide an answer to the first question posed above: Yes, it can well be useful to compile exhaustive lists of genes, proteins and other biomolecules, not as an exercise *per se*, but in order to help understand what goes wrong in our body and to find remedies to put it right.

The result of each examination is more like a photograph rather than a film, given that the diagnostic observation catches just those elements which happen to occur in the sample at the very time at which it was taken or at which the radiograph was obtained. Everything which happened before or after this moment or somewhere else would have been — so to speak — outside our field of observation. Therefore such an examination can also provide false positive or false negative results, depending on the degree of specificity and sensitivity of the test. For all these reasons we reach in many cases only a very approximate understanding of what may be dysfunctional in the patient without exactly knowing how or why the normal function is compromised.

Diagnostic innovations, as inherently incomplete and imperfect as they may be, continue to change the definitions of diseases. The international classification of diseases and related health problems compiled by the World Health Organisation (WHO) is the most authoritative catalogue of disease definitions around. From the time of its birth, more than 150 years ago, it has continued to change.

Each change and each definition reflect a typical medical compromise between body part afflicted and nature and origin of the disease.

These growing definitions which are based on accumulated knowledge and the development of new techniques, generate clues well superior to those available in the past. Above all, they allow the doctor to formulate ideas as to what may underlie a disease, and they may help him or her to imagine how the disease may respond to a known or desirable but still non-existent therapy.

The doctor can identify many variations of a disease in tissues or organs of patients, and the scientist possesses lists of thousands of molecules or diseased cells extracted from the patients' body which he or she has never seen. Only the dialogue between these two worlds, which aim to achieve the same ultimate objectives but use different languages, can succeed in narrowing the field to the most promising molecules worthy of potential development into new therapies.

Without this dialogue the doctor would carry out diagnoses which lack the benefit of the latest molecular insights, and the scientist would compile lists of molecules with dubious value for the patient.

And here is ultimately the reply to the second question: Only the proper extended and fruitful dialogue between medics and scientists will help discriminate between those molecules, among the thousands on the lists generated by omics endeavours, which truly cause diseases and thus might legitimately serve as targets of a new drug, and those which are there by chance as simple bystanders.

Even such fruitful dialogue can fail to engender success, and history provides ample examples of many potential targets which ultimately proved unsuitable for producing useful medicines. For example, cystic fibrosis is a genetic disease for which we have known for at least 25 years the mutated genes and the proteins which cause abnormal secretion in the lungs and some other organs. Irrespective of the most optimistic expectations at the time, a solution has failed to emerge even after a quarter of a century of valiant research. Many other diseases share this property, i.e. ample molecular knowledge contrasting sharply with a desert-like lack of prospective therapeuticants. The road to new drugs is truly arduous!

Is this Target "Druggable"?

Now let us imagine that some medics or scientists have identified a circuit in a given disease which has gone awry, and that in this circuit there is at least one molecule which might serve as target of a yet to be discovered drug. How do we go about the business of designing a new drug which gets there and blocks the target, or at least interferes with its dysfunction?

For a start, we need to understand whether the target is at all vulnerable towards the attack of our future drug. In the pharmacologist's — sometimes pretty awful — jargon a target of this kind is called "druggable".

Screening of Chemicals

Let us imagine that we have investigated in depth the molecular mechanism of a disease, selected a druggable target which seems to play a key role, and decided upon the most suitable formulation and route of administration for the prospective new drug. All that remains to be done now, is to select a compound which works against our target.

For that we have several possibilities. The traditional and arguably fastest approach is to conduct so-called "screening", i.e. systematic evaluation of a large number of compounds present in a particular collection of chemicals, to explore which of these might be the starting point for new drug design.

The idea is simple, but in practice the number of compounds in a typical collection of chemicals owned by a pharmaceutical company can surpass three million. How is it possible to analyse three million compounds in a few days?

The answer is encapsulated in the three words "miniaturisation, robotics, informatics". Let us start with miniaturisation. Collections of compounds tend to be contained in a series of small metallic rectangular boxes of about the size of a postcard. In each box, there are plates with thousands of little wells, and each well holds one compound in a tiny test tube. A collection of three

Figure 1. The drug screening robot at the Mario Negri Institute in Milan, Italy (with the kind permission of Felice de Ceglie, Mario Negri Institute).

million compounds fits therefore comfortably into about 20 boxes and occupies a few shelves of a laboratory cabinet.

Now to the robot (Fig. 1). Robotic arms resembling those used to assemble a car conduct the screening experiments in automatic mode moving the boxes from the cabinet to the area where the test is performed and back again. Each one of the millions of compounds is thus brought into contact with a liquid containing a group of cells or a cell surrogate which reproduces an aspect of the targeted disease. The mixture includes also a substance which releases fluorescent light if the compound produces a desired effect.

These experiments, which would take decades of human manual labour, are completed by the robot in only a few weeks with a precision achievable only by a machine.

To make all this fit together one needs informatics. The actions of the robot which performs the screening are guided by appropriate computer programmes which can also measure the intensity of the fluorescent light emitted every time a compound generates a desired effect. In this way, informatics programmes construct in real time a register of screening results.

Not all compound collections are the same, and some are rather specialised. For example, trabectedin is an antitumour compound which arose from the screening of a collection of marine compounds. In this case, the desired function searched for in the screen was the capacity to kill cancer cells which divide rapidly, a hallmark of cancer.

We shall talk more about trabectedin, the commercial name of which is Yondelis. This drug which was discovered and brought to market only relatively recently serves here as a paradigm of those agents which have made it successfully through all the tortuous phases of drug development, and its history contains lessons worth telling. Now back to drug screening.

If we are fortunate, we may have obtained from screening at least one promising compound warranting further work. Scientists call such a compound a "lead" to indicate that this is a preliminary chemical worthy of being subjected to initial tests. If early experiments on the lead compound furnish encouraging albeit improvable results, its structure can be modified in a way which attenuates or amplifies certain characteristics.

Then, if we know the shape in three dimension of both the lead and the pharmacological target, we may be able to simulate their interactions on the computer. In the pharmacologist's jargon the target is druggable if its molecular surface presents at least one cavity or docking point into which the drug can insert itself. If instead it is totally flat, the target is undruggable.

Next we need to find out whether our target is unique or has siblings with high resemblance.

In biology similar molecules which do similar — although not identical — things tend to be the rule rather than the exception. They are the result of millennia of evolution which have created a sort of intelligent recycling with savings function. An existing and functioning biological molecule can acquire a slightly different structure and altered function consequent to a DNA mutation.

If the altered function happens to be useful to the organism, the similar but different molecule establishes itself as chosen by natural selection.

Examples of such variation by resemblance are the so-called protein kinases, a type of molecular interrupter capable of activating or inactivating other proteins by means of attachment of phosphate moieties. Protein kinases are also highly preferred targets of pharmacologists, because they possess cavities in their structure into which drugs can dock and thus inactivate them. The target of imatinib, which we introduced in Chapter 1, is a protein kinase. Alas, there are at least 500 specimens of kinases with a high degree of similarity! Not surprisingly, commercially available kinase inhibitors inactivate many proteins other than, but similar to, the target ones, thus setting in train unwanted side effects.

Biologicals

Sometimes the screening of chemicals fails to deliver a promising lead. Then one can pursue a more adventurous but also more accident-prone path towards new drugs, taking advantage of molecules already present in nature. Insulin and artemisinin, which we have already talked about, are two examples of this type of drug.

Biological drugs (or "biologicals") owe their name to the "factory" in which they are being produced. This may be the cell of a microorganism, a plant or an animal into which we have been able to insert one or more genes, mostly human ones, with the help of techniques involving recombinant DNA. The adjective "biological" serves to distinguish such agents from compounds generated by artificial chemical synthesis.

However, we need to understand that chemistry is fundamentally involved in each reaction and energy transfer, whether natural or synthetic. And the cell is the most sophisticated chemical laboratory which has ever been invented, as we already discussed earlier. Without chemistry we would have neither biological nor artificially synthesised drugs, and there would also not be water, food or even ourselves.

A commonplace cell into which we insert our gene can generate molecules as sophisticated and perfect as those which our body produces. In contrast, the compounds which come out of — even the best — laboratory of the most modern chemical company are infinitely more simple and unrefined.

However, we cannot control each detail of the biological process as it emerged from the tortuous and random path of evolution and not from a project designed by humans. In contrast, compounds synthesised artificially possess a structure defined and controlled down to the last atom. This is because they are produced according to instructions drawn up by humans. So even though such compounds are so much less sophisticated than agents produced by nature, we do know pretty much everything about them, how they were made and how pure they are.

Let us turn from theory to practice and take Erythropoietin (EPO), a well-known biological, as an example. EPO, a hormone which our body generates to stimulate the growth of red blood cells, which as you may know transport oxygen in the organism. EPO can give a professional cyclist an abundance of red blood cells, thus enabling him or her to speedily negotiate alpine hairpin turns. Or it can cure a patient who has lost red blood cells because of undergoing chemotherapy. The former use is illicit and dangerous, as it is uncontrolled, whilst the latter is legitimate and life-saving when performed under strict medical control.

EPO is a small protein of 166 amino acids containing four branched chains of sugars, the quantity and composition of which is variable and unpredictable.

The vial with EPO available in the pharmacy contains therefore a heterogeneous mixture of molecules. Each of these harbours the same 166 amino acids, but the sugars vary from molecule-to-molecule. Yet these sugars are much more than the proverbial icing on the cake. The stability and activity of EPO depends on the sugars present, even if the individual contribution of each type is unknown (Fig. 2).

Sugars are only one example of what can be added to proteins beyond their synthesis proper. Other chemicals such as fats can also be stuck on to proteins to enable fine-tuning of their properties.

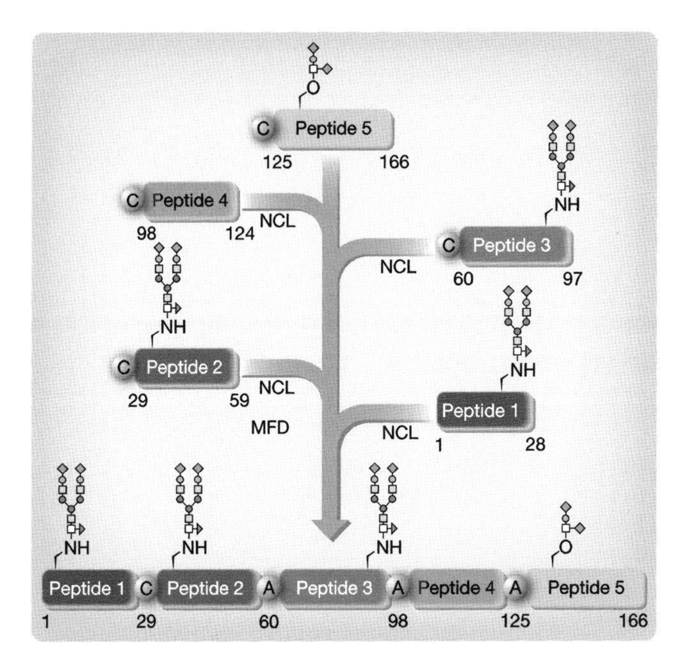

Figure 2. The structure of erythropoietin with its very variable sugars (redrawn from *Science*).

In that way the cell produces molecules of a complexity far superior to that which can be obtained in a chemical laboratory. A small, artificially synthesised compound typically contains less than hundred atoms, in the case of acetylsalicylic acid the number is 21, whilst recombinant insulin produced by a cell consists of 788 atoms organised in 52 molecules. Even more evolved biologicals can contain more than 20 thousand atoms.

The instruction manual which the cell uses to provide the product with often most complicated finishing touches is much less understood and standard than that governing the syntheses of DNA or proteins. Therefore we are still far from being in a position of knowing as to how to reproduce these processes outside a cell.

To design biologicals without cells, reproducing the whole process of development and production in a chemical laboratory would have the advantage of allowing control of their purity and content atom by atom. But we want to stress that in order to do this we would have to be able to manipulate and imitate a cell, and we are a long way from that goal.

So this is why we can expect a biological drug to exert more precise activity than a chemical designed and produced by humans, although we have to settle for less purity and synthetic manipulability down to the atomic level, in the case of the former, which are feasible for the latter.

Molecular Targeted Drugs

Another new drug category which is currently all the rage in pharmacology are the molecular targeted drugs. They are compounds designed to hit a specific target inside the body in contrast to drugs which apparently hit organs rather indiscriminately. For example the so-called monoclonal antibodies are exemplary molecular targeted drugs because they are able to recognise specifically one biological molecule among many. They are also biologicals, as they are obtained by exploitation of a process, i.e. antibody production, which happens to occur in our body.

We encourage you to follow us now without being intimidated by some obscure names, so that we explore together how a monoclonal antibody is produced. We take a mouse and inject our target into it, in other words a substance generated by the human body and unknown to the immune system of the mouse. Therefore the mouse starts producing antibodies against the target and accumulates a stockpile of these antibodies in the cells of the spleen. At this point we take some spleen cells from the mouse and culture them together with cells originating from a tumour called myeloma, which is known to produce a single type of antibody in great quantities. By fusing the cells from the spleen with those from the myeloma, we obtain cellular hybrids which in turn produce great quantities of antibodies directed against our target. These antibodies could be our drug.

An example of a monoclonal antibody in frequent clinical use is trastuzumab, the commercial name of which is herceptin. It attacks a molecule called Her2 present in some breast cancer cells. You can recognise drugs based on monoclonal antibodies by the suffix "mab" (abbreviating "monoclonal antibody") at the end of the name of the molecule.

The considerations discussed above for biological drugs also apply to monoclonal antibodies. The vial available in the pharmacy contains a mixture of more or less identical antibodies. All of them hit the same target, but sugars and other constituents with which each antibody has been decorated are hardly ever identical. A product of this type is on the one hand very efficacious because it has been made by mother nature in a way which we do not know how even to only approximate. On the other hand, it is less pure and its synthesis less controlled when compared to a drug synthesised in the laboratory.

Other examples of molecular targeted drugs are kinase inhibitors such as imatinib which we have already discussed in Chapter 1 and those which block a process called angiogenesis, i.e. the construction of blood vessels, just to name but two types. Generally speaking such inhibitory drugs are recognisable by the suffix "–inib" (standing for "inhibitor") at the end of the name of the molecule.

Fashion, Advertisement and some Thoughts

Are biological and molecular targeted drugs inevitably better than traditional drugs? As we have said before, the passion for anything "bio" extends beyond the supermarket, and doctors and patients are not immune to fashion. Therefore the pharmaceutical industry exploits the interest and curiosity for trendy biologicals, and also more generally for anything technological and advanced, with great and persuasive advertising campaigns. That does not, however, mean that biologicals or molecular targeted drugs are as per definition superior to traditional drugs or that traditional drugs lack a molecular target. Each drug is eventually evaluated for its efficacy and its safety irrespective of its origin — chemical or biological — or whether it has a known target or not.

All drugs, from the oldest to the most recent ones, have at least one molecular target and often more than one. Otherwise they would not work! Similarly, many drugs imitate biological processes or interfere with them, although we fail to understand where and how. We have talked about this earlier when we alluded to agonists and antagonists. In some cases, as for aspirin or imatinib, we do know both, the targets and the mechanisms. In the case of other drugs we know neither, but that does not mean that they lack targets and mechanisms and that they cannot exert activity in a specific fashion.

To say that only new drugs have a precise molecular target does not make sense, because it misguides doctors, patients and journalists. And it demeans the quality of many excellent chemically synthesised drugs which continue to serve their purpose optimally at much more moderate prices than those at which many recently discovered drugs are available.

How to Overwhelm Barriers against Poisons

In order to design our drug, we need also to know where its target is located. If, for example, it is localised in the brain, it would be exceedingly difficult for any medicine to get there. Our precious grey matter, on which the function of each body part depends, is protected by an almost impenetrable entrance barrier. The same holds true for germ cells in which the information as to how to make our heirs is safeguarded.

That after about 200,000 years of human history we still exist and moreover have multiplied to above seven billion, is because we have been blessed with robust protection mechanisms not only in the brain and the germ cells but in all the most delicate parts of the organism. The ability to avoid or deactivate toxic agents is a most potent protection mechanism.

Effectively, evolution has transformed us over time into some sort of fortress. Each entrance leading to the most delicate organs is endowed with a selective barrier, and only molecules which possess a "safety permit" can enter.

The mouth has been the principal point of entry for the major part of human history, before syringes and injections came into being. After having entered the mouth, an ingested substance passes through stomach acids to reach the alkaline environment of the intestinal tract. In spite of the strong digestive treatment of the stomach, a few substances manage to arrive in the intestine intact and sometimes even attempt to pass through the gut wall and to enter the blood stream. The membranes of intestinal cells are lined with special protein pumps, the so-called P-glycoproteins, which are efficient "bouncers" and throw more or less each undesirable substance out. Not by chance these proteins are among the main components responsible for the resistance of the body against drugs. The resistance against antibiotics, mentioned in Chapter 1, is often mediated by these pumps.

To deal with the molecules which distract the intestinal bouncers, there are veritable "bomb disposal experts" which can defuse each potential "bomb" rendering it soluble, so that it can be eliminated with the urine or the faeces. The body's main elimination task is carried out by the kidneys, the body's principal garbage filter, which expels most rigorously all aqueous waste. For some drugs, though, generally the ones of high molecular weight, the principal elimination path is into the faeces via the bile which is produced in the liver and flows into the intestinal tract.

Now you have learnt how our body's antipoison headquarters constitute a tremendous obstacle for a drug to overcome, considering it is a foreign substance which our organism recognises as potentially poisonous. The same is true for each new substance which the history of human evolution has yet to include in the catalogue of harmless substances.

Poison or Remedy? Depends on the Dose

"All things are poisons, for there is nothing without poisonous qualities. It is only the dose which makes a thing a poison" wrote the Swiss alchemist Paracelsus in 1538, and his most famous adage is still valid today (Fig. 3).

Even a harmless substance such as water can be toxic after several litres ingested within an hour or so. But the contrary is also true: Potently poisonous alkaloids, products of some plants, have therapeutic effects when used at the right doses. Examples are the

Figure 3. The portrait of Paracelsus conserved at the Louvre Museum in Paris.

Figure 4. Periwinkle is an ornamental plant, originally from Madagascar, from which the anticancer drug vinblastine is isolated (Biswarup Ganguly, photo reproduced under Creative Commons License).

alkaloids extracted from the *Vinca* (periwinkle) plant, which at therapeutic doses possess potent antitumour activity (Fig. 4).

In nature, poisons are more the rule rather than the exception, because toxins confer a considerable evolutionary advantage on plants, mushrooms or generally all organisms fixed to the soil, where they cannot escape attacks from predators. A harmless plant or mushroom faces the high probability of being devoured by some animal, unless it is covered with thorns or repellent.

Poisons are common also in some marine animals, such as tunicates, which live attached to rocks. As soon as they are approached by a predator they emit their toxins into the surrounding water. One of the poisons produced by tunicates has become trabectedin, a drug which we have already introduced, and which at the correct doses exerts beneficial activity against sarcomas, i.e. soft tissue tumours and ovarian cancers.

Poisons in the aquatic world are generally more potent than those from plants or mushrooms, because they need to function when diluted in water. Not by chance, one of the most powerful poisons overall, tetrodotoxin, is produced by the pufferfish. An antidote against this deadly toxin does not exist, unlike what fiction such as 007 seems to suggest.

For pharmacologists who consider the world of natural products an exciting hunting ground for new drugs, their dose-dependent ambivalent character — harmful or beneficial — is an extraordinary source of wonderment and inspiration.

The chemical structures of natural products which have established themselves in the course of evolution are extremely complex, refined and difficult to imitate. Some chemists consider success in the reproduction of such compounds with most elaborate structures a stimulating intellectual challenge.

You may ask yourselves why we should replicate a substance if nature on its own is capable of producing it so competently. Sometimes a natural product has a desirable therapeutic effect, but it may also possess toxicity. If its laboratory synthesis was possible, one could eliminate the portion of the molecule which renders it unsafe or substitute this with a molecular feature which makes the compound less harmful. This is what happened, for example, in the case of salicylic acid with the addition of an acetyl group. A small modification of its chemical structure generated efficacious and safe aspirin.

Another reason for replacing a natural product with a synthetic one is that nature can be a rather unreliable factory. The harvest of a plant often depends on season and weather. Also the concentration of the desired constituent may vary from plant to plant. For example, if a plant is exposed to predators it may face strong stimulation to produce more of a toxin than one which does not face this challenge.

Artemisinin, for example, is a drug the production of which has passed from nature to biotechnology, in order to avoid fluctuations in harvest and price. It is extracted from a shrub called *Artemisia annua* (Fig. 5), and it is one of the most commonly used drugs against malaria. The effects of artemisinin against malarial fever have been known in China as early as in 340 AD, but the active principle of Artemisia has been isolated from its leaves only in 1972.

The gene necessary to make artemisinin was inserted into a microorganism only recently. It can produce the drug in a much more constant, predictable and economic fashion, compared with production in the plant with inevitable variation in harvest and price. This is of course excellent news for the more than 200 million malaria patients.

Figure 5. The *Artemisia annua*, from which we extract artemisinin, one of the most effective drugs against malaria (Jorge Ferreira/Wikipedia).

There are many other naturally occurring substances turned useful drugs. Among them are alkaloids such as morphine and codeine extracted from the opium poppy plant, most useful in pain therapy but also abused as recreational drugs. Then there are scopolamine, obtained from the belladonna plant, which has helped generations of mariners and astronauts to combat motion sickness, and paclitaxel, an antitumour agent extracted from the poisonous bark of the Pacific yew tree.

Routes and Formulations

To design an efficacious drug, we must also know the path which it may take until it reaches the target. In order to optimise this, we need to establish its most adequate formulation.

The oral route, via the mouth (or "*po*", from the latin *per os*, "by mouth"), is the most convenient. This is especially the case for the treatment of chronic diseases which require daily administration for life, probably several times a day, of the same medicine, either in the form of pills, drops or syrups.

In contrast, intravenous or intramuscular injections are inconvenient, because they require somebody to stick the needle in. Yet, injections have to be used in the case of drugs of limited stability or for fine tuning of a dose.

In the case of the intradermal route, a small plastic container is implanted below the skin of an arm or the stomach, and it releases regular doses of the drug at preset intervals. This arrangement is used for medicines, the administration of which needs to be regular and must not be forgotten, as in the case of insulin against diabetes or of contraceptives. For the control of diabetes there exists an intelligent intradermal pump capable of interacting with a monitor which measures blood glucose concentration and as a result calculates the most appropriate insulin dose. Such regular calculation of the right dose after monitoring of the glucose concentration avoids potential insulin over- or under-dosing (Fig. 6).

Finally there is the respiratory route for drugs administered as aerosols, used for example for anaesthetics to sedate a patient before an operation or for antiasthma drugs.

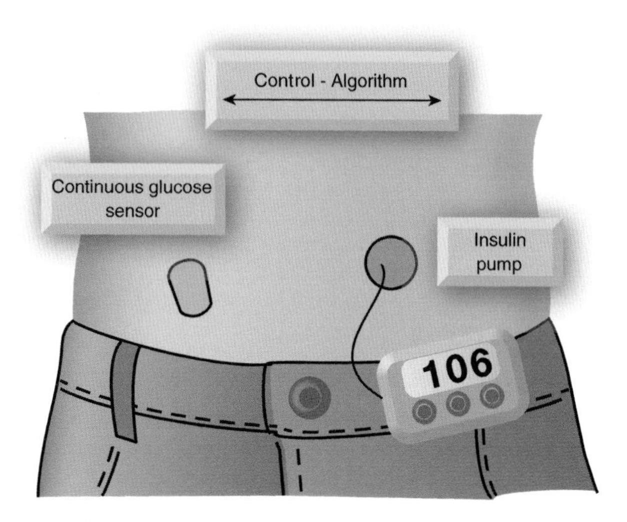

Figure 6. An insulin intradermal pump connected to a blood glucose monitor (redrawn from: University of Illinois).

Each administration route and formulation has advantages and disadvantages, so it is incorrect to say that one is *per se* better than another. The choice depends on the health problem which needs to be treated and on the properties of the drug. For example, a pill is convenient to be taken, but its ingredient drug has to pass through the acidic stomach and then the alkaline intestine. In order to get through these two oppositely charged environments unchanged, a drug molecule needs to be neutral and non-polar, i.e. lacking an electronic charge. Not all drugs are absorbed and delivered to the target in the same way, and the characteristics of these processes depend in part on the intestinal barrier and on liver metabolism. Importantly, there is considerable variation in these processes between individual patients. In the case of medicines taken by mouth one needs also to take their taste into account, which should not be unpleasant.

For drugs which have to be injected, the active principle should be soluble in water or in a solution with characteristics similar to those of our body fluids. If this is impossible, substances have to be added which increase drug solubility. Paclitaxel, for example, is an antitumour drug characterised by very low water solubility. It is therefore sometimes combined with castor oil, in which paclitaxel is soluble. But at high doses this combination can evoke unwanted side effects.

So the formulation of a drug has to make sure that the active principle does not decompose and that it is able to reach the correct part of the body at the right rate and concentration.

And now — Do we Have a New Drug?

Our new drug may be a small compound synthesised in the chemical laboratory, a natural product or the product of a modern rational drug design project. It may be a remedy at a certain dose and a poison at another higher dose. It may pass through the stomach and the intestine or enter the blood circulation directly. But all these possibilities do not tell us whether the drug will eventually really work and be safe. In order to find that out, we need to embark on intense investigations. In the next two chapters we shall see how one performs experiments on a new drug molecule from the early stages of laboratory research to studies in animals and in human beings.

Chapter 3
Laboratory Studies

The human body is amazing. We think that we know all of its nooks and crannies, and then a new ligament crops up. In 2013, a group of Belgian scientists actually discovered one close to the knee. You can imagine the reaction: "How is it possible that a ligament has remained undetected by generations of orthopaedic surgeons, anatomists, pathologists, physiotherapists and sport physicians?" Are we not, after all, in the era of magnetic resonance and of the transparent body? These are legitimate points as long as one avoids looking into a human body, a wet and dark mass which is very difficult to reconcile with the neat and clean images presented in anatomy atlases.

We Are Not All the Same — And Mostly Do Not Know Why That Is

Think of your friends' faces. They have more or less the same bits: a nose, two eyes, a mouth, two ears, a front and often some hair. It is, however, the way in which these elements are combined together with traces left by scars, diseases and age that result in the unique face each of us have at each stage of our lives, even in the case of identical twins. So it should not surprise us too much that a big piece like a ligament has been discovered only in 2013. And the rule of diversity in similarity holds also for the small bits and pieces which we are made up of.

Genes and proteins and also cells, tissues and organs share a general architecture which has been conserved in most animals. After all, we are the result of a common evolution. Many details vary though, and it is the details which make the difference as far as diseases are concerned. They determine the way in which diseases present and the diverse responses to medical intervention.

The study of potential responses to a drug starts in cells, or rather in a part isolated from the organism in which it is possible to mimic — at least in part — the disease under study, and the way in which we intend to cure it. And this is the very first phase of investigation which takes place in the laboratory, called "preclinical" because it precedes the clinical trial of the drug in patients. Only if the results of the initial tests in the laboratory are positive can we proceed with the successive phases of experimentation.

Experiments Using Cells

In general, the testing of a drug in the laboratory involves exposure to the drug of diseased cells obtained previously from patients by means of a blood sample or biopsy, a surgically removed tissue fragment which can help with the diagnosis.

The cells need to be capable of growing outside the organism in a plastic flask, which contains a liquid with nutrients called "culture medium". Not all cells obtained from patients survive under such artificial conditions. In some experiments one has to use other cells, sufficiently similar to those from patients but capable to survive in culture.

The drug under test is added to the culture medium containing cells to evaluate and observe whether the problem disappears or is reduced to an acceptable level. If, for example, we evaluate a potential antitumour drug, we want to see that tumour cells exposed to the agent die or stop growing. Cancer pharmacologists say — somewhat tongue in cheek — that the only tumour cell they like is a dead one. If however we want to test a drug against a degenerative disease, in which diseased cells die sooner than they should, we need to verify whether the drug can prolong the life of these cells.

If on exposure to the drug the cells stop proliferating, die or reacquire their strength, these effects can be observed under the microscope. Biochemical or genetic experiments can measure subtle actions of the drug, such as the change in concentration of certain proteins or the modification of a gene sequence.

In addition to experiments on diseased cells, control experiments need to be conducted, in which healthy cells of a makeup similar to that of the diseased ones, are exposed to the drug. If the drug acts specifically against the disease, the cells in the control experiment should not show any effect. Additional controls are required in which the diseased and healthy cells are subjected to experimental manipulations under identical conditions, except that they are not exposed to the drug. This will help rule out that experimental factors other than the drug influence the results and confuse the scientists.

Importantly, results of experiments with cells are only indicative, and they have considerable limitations. Cells which grow well in the laboratory, may not really be representative of the diversity of the cells contained in the diseased tissue. Furthermore, under artificial and isolated laboratory conditions, cells in culture do not receive any signals from other cells which could influence cellular behaviour, as they would do as part of the intact organism. Also, cells grown and propagated in the lab for many years tend to develop characteristics which differ dramatically from those of the tissues from which the cells originate. For example, the sixty cell lines which scientists at the National Cancer Institute in the US have derived over the years from tumours from various cancer patients should represent the majority of human cancers. These cells serve often as experimental system in which initial tests of novel potential antitumour agents are conducted, even though these cells reflect only to a small degree the immense variety of tumour cells which are found in patients.

So in experiments in cells we can obtain indicative results, but we cannot use the results to predict the response of a complete organism to a new drug. In order to be able to do that, so-called *in vivo* tests are necessary, that is studies in whole organisms which develop an illness similar to, although perhaps not identical with, that in human patients. We are talking about experiments in animals.

Why Are Experiments in Animals Necessary to Evaluate a Drug?

"I, Caterina S, am 25 years of age thanks to responsible research including animal experimentation. Without this research I would have died aged 9. You have given me the gift of a future."

Around Christmas 2013, these words went across a social network. They were written by Caterina Simonsen, a girl with several serious illnesses, who owes her life to medical research. As soon as they were online, they immediately sparked violent reactions, threats and cruel insults.

For the people who insulted Caterina, experiments of drugs on animals are unacceptable to the point that they accept the death of human beings. But there are many people, neither fanatics nor monsters, who are uncomfortable with the idea that tests using laboratory animals are indispensable to prove the safety and efficacy of a new drug.

Serious reasons support the unavoidable link between animals and research into new drugs, the prime reason being drug safety.

Some effects of a drug become visible only in a complete organism, equipped with all the organs which the drug can reach and be affected by. These effects would not be detected in isolated cells which are used in experiments *in vitro*.

Although it is true that we humans are in many respects different from animals, we need to remember that with some animals, we share a great part of the evolutionary path and therefore share with them molecular circuits which have been conserved up to this day. Cells, for example, are organised in almost identical fashion between all mammals. Even more striking is the fact that many constituents are virtually interchangeable. It is rare that a human protein would not function at least to some extent when it substitutes its counterpart in a mouse.

So we living beings are more or less strictly related, and so are our molecules. By virtue of this "molecular kinship", the results of experiments obtained in animals sufficiently similar to us can give us useful indications. They can, for example, suggest that a drug is tolerated by both humans and animals, that it does not pose problems to the heart or the respiratory system, that it does not provoke sedative or stimulatory effects, or that it fails to modify hormonal equilibria and so on.

For these reasons, experiments in animals are a source of most useful information for pharmacologists. And in many countries, laws request that experiments on a new drug are conducted in more than one animal species before one can proceed to the clinical experimentation in humans. These laws reflects the demand of citizens, especially in Western societies, for drugs which are prescribed by the doctor and approved by medical regulatory agencies to be safe.

We not only expect safety and absence of side effects of a new drug, but tend to take them for granted. And these expectations require experimentation on new potential drugs to be done in laboratory animals. Experiments with this goal are regulated by strict guidelines protecting the well-being of animals and limiting their use to the absolutely minimum necessary.

The European Union (EU), for example, imposes the so-called rule of the three "R"s on projects financially supported by the Community. The three Rs are firstly *replacement*, refraining from the use of animals if equivalent alternative methods exist, secondly *reduction*, using the minimal number of animals possible, and thirdly *refinement*, assuring the best quality of life for the animals and the least amount of suffering possible.

Researchers do stick scrupulously to these rules. Moreover, many of them would gladly do without tests in animals, if this part of the experimental plan was not first and foremost required by law, vitally safeguarding the safety and efficacy of new drugs.

A Model for (Almost) any Disease

Researchers call each living non-human organism an "animal model", if it is useful in the study of a disease or its cure. The goal is to understand as many aspects as possible about the illness itself and to develop the foundations for a cure, always mindful of avoiding harm to people caused by premature clinical experiments which would be considered unethical.

The most frequently used laboratory animals are rodents such as mice and rats, which are similar to us with respect to both genetic characteristics and many of the diseases which they develop. But also creatures such as fruit flies, worms and fish can be useful in the study of molecular circuits, which although present in rodents and mammals, are more experimentally accessible and therefore more easy to analyse in these relatively less complex animals (Fig. 1).

The diseases to be studied in animals can either occur spontaneously or be induced. For example, one can reproduce human infectious diseases by exposing animals to bacteria or viruses, causing the same infections as in humans. Or one can induce some type of tumour with carcinogens, for example, cigarette smoke, which causes lung cancer also in animals. Or it is possible to modify the DNA of animals because that modification reproduces a human disease most faithfully.

Figure 1. The zebrafish (*Danio rerio*) is a commonly used animal model in biomedical research (Azul/Wikipedia).

A technique used especially to study potential antitumour drugs entails so-called tumour xenografts. These are pieces of tumour obtained from patients and injected into mice which fail to reject them because their immune system is blocked. Then the drug under study is administered to the animal. If the transplanted tumour decreases in size or disappears, whilst the tumour in the control animals which did not receive the drug is not affected, the drug has had a positive effect.

Unfortunately, the tumour which grows in these animals does not always reflect what happens in the patient's disease. The biological differences between humans and animals in diet and in habits which can influence cancer development are enormous, and obviously there remains still a lot of research to be done in order for us to be able to imitate in a precise way a human disease in an animal. Even with these limitations, human tumour xenografts are an experimental model which is far superior to cells *in vitro*. And xenografts are among the best experimental tools hitherto available which allow the testing of new drugs before they move into human experimentation. In some very advanced hospitals, xenografts are used to help with choosing those drugs from among the available ones, to which the patient's tumour is most sensitive. This method is still at an experimental stage.

In order to be really useful, animal models need to represent a disease of which origin and mechanisms are well known and also similar to the equivalent human disease. But many diseases are complex and difficult to model in an animal organism. Think for example of autism, schizophrenia or other cognitive disorders of the nervous system, of which we understand very little in terms of their causes and how they develop in human beings. It is difficult to envisage that diseases of such immense complexity could arise in animals lacking our language or articulate thinking abilities. Nevertheless, there are many attempts afoot to generate reliable animal models, if not of the entire spectrum of the symptoms of such diseases, at least of some of their details. And if only a few such disorders were to be ameliorated in animals, due to treatment with an experimental drug, who knows, this drug might cause treatment improvement also in humans.

Toxic or Harmless?

Tests in animals may eventually discover a potential new drug to be toxic to the liver, kidneys, intestine, blood, or other organs. Such effects may not be the same in humans, but they would certainly not have been observed if the drug was only studied in cells *in vitro*.

An investigation in the year 2000 has estimated that with the help of animal models, one obtains results on drug toxicity which in 71% of cases are consistent with those seen in humans. Research involving animals are therefore especially useful to provide indications as to whether a drug may cause toxicity in humans. Drug safety evaluation is indispensable at this moment in time, in which unfortunately, as far as we know, medicines with both high efficacy and absolute safety just do not exist.

Drug safety tests are well laid down by law, with respect to both laboratory practices and current regulations. Potentially vulnerable organs from the animals in which the new drug may be toxic are painstakingly analysed to establish the harmful dose and the time after administration at which such an effect may occur, and whether it is reversible or cumulative.

How the Organism Modifies a Drug: Pharmacokinetics

Let us imagine we know what to do against the disease under study and that we have an ideal drug. We know to the last detail the circuit which has gone haywire, and we have designed a drug which aligns perfectly with the dysfunctional molecule, or inhibits it, and studies in cells *in vitro* have furnished encouraging results. We expect all to proceed well in subsequent early studies in animals and then in human beings. Well, that is unless... the animals' liver transforms the drug rapidly into an inactive substance, or the kidneys eliminate it in just a few minutes, so rapidly that the drug cannot exert its effect. Or the agent remains in the organism for too long a time and thus causes toxicity.

A drug gets first into the blood, after that it penetrates the organs, and then it is altered by the liver, so that it can be more easily eliminated. Elimination can occur in the liver via the bile into the faeces, or through the kidneys into the urine. The time it takes for a drug to be distributed in the body and eliminated from it, is assessed by the so-called pharmacokinetics. This is a branch of pharmacology which studies how the concentration of a drug in the organism changes with time. In simple terms, pharmacokinetics describes what the organism does to a drug, how the drug is absorbed, into which organs it is distributed and how it is transformed and eliminated.

Let us briefly dwell on the topic of transformation, as it is crucial for the understanding of how the body interacts with drugs. Each substance is potentially subjected to alterations of its chemical structure when it passes through the liver. This is the case not only for drugs but also foodstuffs. The process of such transformation is often called metabolism, and its aim is to render a substance easy to eliminate with the urine or faeces. There are enzymes, particularly in the liver, capable of modifying chemicals and generating secondary products or metabolites. Metabolites can be as active as, more active or less active than, the parent drug. The same holds true for the side effects.

The nature and extent of metabolism differ from compound to compound, but also from person to person, and they are not the same in animals as in humans. Pharmacokinetic curves obtained in the laboratory can sometimes anticipate, in an approximate fashion, potential problems which the researchers can try to overcome or avoid in humans. For example, a very active drug, however poorly tolerated by the kidneys, should be administered to patients with renal problems only if it really offers an overwhelming therapeutic advantage.

How the Drug Alters the Organism: Pharmacodynamics

Apart from what the organism does to a compound, one can start to measure in animals also what a drug may do to the organism, in other words the response of tissues and cells to the drug. This is called pharmacodynamics.

The Therapeutic Index

Out of the whole of the studies of efficacy, toxicity, pharmacokinetics and pharmacodynamics emerges the so-called therapeutic index. This is the measure of how a treatment affects the diseased tissue in terms of activity, as compared to the entirety of its undesired effects against normal tissues. Ideally, each useful drug should have a high therapeutic index, i.e. it should be highly efficacious and safe. In reality, it is difficult to reach a high therapeutic index, because the majority of proteins of a diseased cell function also in a healthy cell. In other words, even a drug with the best therapeutic index is never 100% selective, and therefore it inevitably causes some side effects.

Towards the First Tests in Humans: Does the Traffic Light Display Green, Red or Yellow?

You may recall trabectedin. Most promising results emerged from studies *in vitro* of this drug obtained from the seabed. It killed a great portion of tumour cells of the sixty cell line panel held by the National Cancer Institute (NCI) in the US; and in studies in mice bearing human tumour xenografts, it eradicated tumours which were resistant against treatment with many conventional drugs.

However — and in pharmacology there seems to be almost always a "however" — these studies in animals showed a toxic effect on the liver of mice and rats, whereupon the NCI, the US partner of the European study team, wanted to terminate any further preclinical and clinical experimentation on this compound immediately.

The European team, however, saw the situation in a somewhat different light. The efficacy data of the drug had been far too exciting to outright reject trabectedin from consideration for clinical trials in humans. So they decided to go it alone, hoping that the problem with the liver may ultimately be resolved.

By proceeding to the clinical evaluation of trabectedin, these oncologists took on a great responsibility. But accepting this responsibility was subsequently rewarded by the results, given that patients today have an efficacious drug at their disposal. And to resolve the problem of the liver toxicity it suffices to treat patients first with an antiinflammatory drug as dexamethasone.

How was this solution found? The problem caused by trabectedin in the liver was found to be preceded by a severe inflammation of the bile duct via which the drug is eliminated. With the aim of reducing this inflammation, researchers tried out dexamethasone pretreatment in the animals. And they observed that the livers of these animals tolerated trabectedin much better than those which did not receive dexamethasone. The results in rats were subsequently confirmed in patients in a clinical trial. Today, patients all over the world receive pretreatment with dexamethasone before they receive trabectedin.

So for trabectedin, the traffic light on the path towards clinical evaluation in humans signalled yellow veering towards red. This type of outcome is much more common than clear-cut red or green, and it poses difficult ethical questions for the pharmacologists and medical doctors who have to take these decisions. Should they stop experimentation for safety reasons, or should they go on to do a trial in humans, attempting on the way to improve the drug's safety? Such dilemmas strongly influence the probability of a drug encountering the cancel button on its passage from the preclinical to the clinical study phase.

Statistics divulge that only one in a thousand compounds gets through the stage of preclinical examination in cells and animals.

This disappointingly small number tells us how difficult the decision for researchers and pharmaceutical companies is as to whether or not to give the go ahead for trying out the drug in human beings. And this difficulty is compounded by the fact that a positive decision necessitates an investment of several billion pounds.

Chapter 4
First Time in Humans

In the middle of the 18th century, the British Empire was at the height of its power. Hundreds of ships of His Majesty's Navy sailed proudly across the oceans of the whole world. The ability to conquer new territories and to stash away large amounts of merchandise and precious objects in George II's treasure chests depended crucially on the voyages of these vessels. But not everything goes always to plan.

In 1747, James Lind, a naval medical officer on board of HMS Salisbury, helped to cope with the umpteenth outbreak of an epidemic of scurvy, an illness which seemed inevitable among crew members of ships after several weeks on board when the sailors started to waste away and to lose blood.

The problem affected more than 12,000 men, that is, all sailors of the Royal Navy. When these men were ill, it was difficult to win wars and bring treasures back home. This estimate does not even take the personnel into account who served on innumerable civilian geographic or scientific expeditions.

So, Dr Lind decided to conduct an experiment. He chose 12 sick sailors and divided them into six pairs. Each sailor on the study had to adhere to a fixed diet, which was the same for all of them. Moreover each sailor had to take a supplement which varied between pairs. The supplements chosen by Dr Lind were substances which, popular belief at the time considered potential remedies against scurvy: cider,

Figure 1. James Lind testing the treatment for scurvy on sailors (James Thom, originally from *A History of Medicine in Pictures*, published by Parke, Davis & Co. in 1960).

lemon, orange, salt water, vinegar or a mixture of garlic, mustard and horseradish.

After six days of treatment, only the four sailors who had ingested lemons or oranges improved, so that they could stay in service. The others deteriorated further (Fig. 1).

The First Controlled Clinical Trial

When the British Navy decided that citrus fruits needed to be part of all sailors' rations, which happened about 40 years after Lind's experiment, scurvy disappeared from the ships of the Royal Navy. This time interval seems rather long, but innovation has always been slow in being implemented. Also, today the time period between the discovery of a possible solution to a health problem and its practical implementation in patients is at least 10 years and at times half a century.

We know now, that the cause of scurvy is the deficiency of ascorbic acid, vitamin C, which occurs in abundance in citrus and other fresh fruits and vegetables. Before this experiment proved how much they were indispensable, fruits and vegetables were absent from the naval pantry because they are highly perishable.

The proof of Dr Lind is considered the first clinical trial in history, because he could establish with clarity and simplicity some fundamental principles applicable to all clinical studies. The environment, that is the general conditions of the trial, need to be controlled and equal for all participants. In the case of Dr Lind's experiment, the diet was fixed. The treatments need to vary between groups. In the case of Lind, each sailor pair received a different supplement. The response of the organism to the different treatments must be interpretable in a clear and unambiguous fashion, that means there either is a response or there is not.

Today, it is difficult to obtain such clean and tidy results because diseases caused by a single external factor are not easy to pinpoint and manipulate, unlike vitamin C against scurvy, which were rare already then, have all by now been explored and resolved. The majority of illnesses which remain to be cured are more complicated ones. They tend to have multiple causes, and their origin can be outside or inside the organism. The nature and combination of causes often vary from patient-to-patient. The response to drugs is very variable. All of these issues suggest that results of a clinical experiment are often unclear, obscure or incomprehensible to the medical doctor or pharmacologist.

In this chapter, we shall see how a modern clinical experiment is set up and conducted, in which a drug previously studied in the laboratory is for the first time explored in humans. Remember the general principles established in Dr Lind's experiment. These, together with modern medical methods, proper controls and the care and foresight which society requests from drug producers, should protect trial volunteers and ultimately lead to safe products.

The Participants

Who are the persons taking part in a clinical study, and how are they chosen? Under ideal conditions, a drug should be evaluated in a population as similar as possible to the people for whom it would be ultimately prescribed. Should the drug show efficacy, it would receive regulatory approval. If, for example, we were to study a drug against osteoporosis, a bone disease which occurs especially in women aged sixty or above, we should include in the study older female individuals rather than young ones. If, however, the aim is to study a drug against diabetes, we would have to include overweight men and women from the age of twenty upwards, with a majority of individuals being over 65, because these are the characteristics of the population in which the disease occurs.

In practice, groups of individuals who take part in studies of new drugs cannot be fully representative of the immense variety of humans and of their enormous biological diversity. Therefore, results obtained from such studies are never totally applicable to each and every patient who would use the drug in real life.

Nowadays, the majority of new drugs undergo investigation in a population of mostly younger people. The younger population tends to suffer from fewer health complications, and has been exposed to fewer treatments, than older persons, who often use drugs regularly. Predominantly healthy male adults are recruited into trials. Women of fertile age, children and senior citizens over the age of 65 are considered too vulnerable and therefore tend to be excluded from trials. But such exclusion is a two-edged sword because drugs are often prescribed particularly for these categories of individuals.

In the light of the increase in human life expectancy, postponement by 10 years of the age limit of volunteers able to take part in trials is currently under consideration. Inclusion of people over the age of 65 is difficult, because there may be age-related health problems such as compromised kidney function, presence of concomitant diseases or the need to take other drugs at the same time, all of which might complicate the trial and confound the results. Yet, in real life about half of all drugs are being used in old age, and therefore it is important that they are evaluated in terms of how they behave in the aged organism.

Children are only included in trials of those drugs which are developed for use in paediatric medicine, and study details are adjusted to the appropriate age range. Approximately half of the drugs prescribed for children have been evaluated only in adults. But, children are not "small adults" for whom it would be sufficient just to reduce the drug dosage in proportion to body weight. For drugs designed for the use in paediatric practice and evaluated in children one has to take into account the fact that the activity of the enzymes responsible for drug metabolism can be different to that in adults, either higher or lower.

In addition to these general criteria which are valid for all clinical studies, the medical doctors responsible for a study have to establish also specific characteristics which the volunteers need to possess in order to qualify for taking part. Together with their staff they try to recruit such individuals and enrol them into the study. It is important to underline here that the persons who take part in any clinical study are strictly volunteers, and nobody can be forced to take part against his or her will. This regulation is part of the Nuremberg Code, a series of ethics principles established in 1947, after the Nuremberg Trial of Third Reich war criminals brought to light the cruelty and abuse committed on inmates of concentration camps in experiments which, by the way, lacked any medical value and scientific rigour.

Informed Consent

"My name is Mary Smith, I work in the General Hospital. In this hospital, we are conducting research into the disease X, which is most common in this country. I would like to invite you to consider participation in this research. Today I let you have some information which should help you to come to an informed decision and to either accept or refute our invitation. I want to stress that you need not decide today. Before you make your decision you need to think about it carefully, and if you require more information I shall be happy to provide it to you."

This is more or less the way in which the conversation would be initiated in which a potential volunteer obtains information on the nature of the clinical trial. This includes the aim of the study, the names of the medical doctors responsible and the hospital in which it is carried out, the type of intervention envisaged, the fact that participation is purely voluntary, information about the drug under study and its potential side effects, and the criteria used to exclude volunteers.

In each serious clinical study, the study team dedicates a lot of time on these conversations, at the end of which the volunteer may wish to sign the so-called *informed consent* form. This document summarises the available information, and each volunteer needs to sign it if he or she decides to take part in the study.

Provision of information in a responsible and comprehensive fashion to volunteers who are to participate is strictly obligatory by law. It is also very important to ensure that the volunteers are in the know to avoid simple errors to occur.

Randomised, Controlled, Blinded — Three Fundamental Rules of a Rigorous Clinical Study

After the recruitment of the volunteers into the trial, the study needs to be organised according to certain rules. These rules are designed to reduce to an absolute minimum possible, statistical errors or distortions caused by psychological influences on volunteers, trial doctors and the health team, and they may well be unconscious ones.

(1) The volunteers are divided in random fashion into at least two groups or "arms". The experimental group receives the drug under study, whilst the other group, the control group, receives

an inert substance (the so-called placebo) or a treatment which is in standard use for the condition to be treated.

Volunteers are randomly assigned to the experimental or control groups, a process called "randomisation", and ideally the volunteers should be distributed to groups by drawing straws. Nowadays, randomisation is performed by computer programmes which assign to each volunteer a number, generated by chance and assigned by chance to one of the study arms.

Ending up by chance in one of the two arms can generate discomfort in patients who expect that one of the two treatments is better than the other, and that he or she should be treated with the better one. Given that such a difference is only divulged at the end of the study, this is never known at the start of the study. If it was known, the assignation by chance of patients into the two arms would create a group A receiving treatment with the better compound, and a group B whose recruits receive the inferior compound. This would not only be unethical, but above all it would not make sense at all. The study is conducted at a point in time when one does not know which treatment is better. In order to avoid problems arising from patients' expectations associated with taking part in the study, it is essential to explain with utmost care the aim of the experiment and everything which is known and indeed not known about the two treatments. Otherwise, the results could conceivably be confounded by an erroneous perception by the volunteers of the value of the two treatments.

(2) The participants in the study are followed up in an identical and controlled fashion. All of them undergo the same tests, receive the same medical visits and undergo the same examinations, whichever group they belong to. The groups differ only in terms of the treatments which the study tries to compare.

(3) The volunteers do not know whether they receive the experimental drug, the placebo or standard treatment. Therefore the study is carried out *blinded* or *blind*. When neither the medical doctor nor the patients know who receives which treatment,

the study is called *double blind*. Studies are called *triple blind* when in addition to patients and doctors those individuals who analyse the results do not know which treatment any of the volunteers has received. In this case, the treatments which the two groups have been subjected to are disclosed only at the end of the evaluation of the results. Only then does one find out if the experimental drug has been efficacious or not.

Placebo and Nocebo Effects

In common language, the word placebo tends to be associated with something useless or superfluous. In real life, the somewhat mysterious phenomenon called *placebo effect* is anything but useless. Sick individuals can react in a positive fashion to a medicine irrespective of the content, just because of the fact that they are occupied dealing with their problem.

Placebo effects, the most studied of which are related to pain, are real and measurable in the laboratory or hospital. They can be attributed to the action of known molecules in the brain, and happen just as likely in the presence as in the absence of an inert substance. Words can suffice to induce a placebo effect!

Often, however, abnormal conditions such as pain can resolve unassisted, and therefore it is sometimes difficult to distinguish between the natural evolution occurring in our body and a potential placebo effect.

To compare a drug with a placebo can yield the definitive result concerning the potential activity of the drug. If the drug's activity is equal to that of the placebo, it means that its active principle fails to produce an effect different to that of an inert substance. It is, however, not always correct or ethical to compare the activity of a drug only with that of a placebo. If for example there is a treatment available for the disease under study, potential differences in efficacy are measured with respect to the already available drug, and it would be unethical to offer only a placebo to the patients in the control arm.

The negative mirror image of the placebo effect is the so-called *nocebo* effect, a harmful effect not ascribed to the action of the drug but to the patient's psychological imagination of the drug's side effects. Sometimes too detailed information on potential adverse effects of a drug can stimulate a nocebo effect, provoking the appearance of symptoms of which the patient has heard. It is perhaps better not to linger too long reading the list of adverse effects in the information leaflets accompanying drugs!

The Four Phases of a Clinical Study

Imagine entering into a large hospital where a clinical study is initiated, a study in which a new drug is tried out for the first time in humans.

The moment is important and emotional for all involved. For the researchers who have worked on that drug for decades and finally should find out whether it works in humans or not; for the medical doctors who hope to be able to offer a new therapeutic opportunity to patients who currently have few; and for the patients and their families for whom the new drug — if it does indeed work — can make a big difference by saving or prolonging the life of a father, mother, son or daughter.

As the emotional involvement with the study can be considerable, it is most important that everybody concerned follows scrupulously the rules governing human trials as described below, in order to avoid any bias or undue influence during the conduct of the study and its analysis.

So let us not just yet enter our hospital, before having recalled some rules of good practice relating to such studies.

Firstly, it is not allowed to draw premature and rapid conclusions. Instead rigour and extreme caution are obligatory. The minds of the researcher and doctor must be prepared to accept any results conceivable, and not only those which they would like or not like to observe. One needs to be utterly impartial, and one has to extract as much information as possible from the emerging results with regard to the properties of the disease and the drug. The scientific curiosity of the pharmacologist must be insatiable!

And whilst the researchers in the trial team have to be rigorously objective and very curious, they must also be extremely cautious and respect one of the fundamental principles which students learn when they first enter the medical faculty of the university: "Above all you must not harm the patient".

The conduct of human trials of new drugs can be divided into four phases, three occurring before regulatory approval of the drug and one afterwards, according to the following Table 1.

In this chapter, we shall discuss the first three phases before drug approval, and deal with the fourth phase in Chapter 7.

Phase one

As the new agent which we want to study has never been tried out in humans, the aim of phase one is to evaluate its safety in a small number of persons, establishing the optimal dose and the best way of administering it.

One starts with two small groups of about 10 volunteers each. One group is made up of healthy individuals, the other of patients.

Table 1: The four phases of the conduct of human trials of new drugs.

Phase	Goal	Typical number of subjects
1.	Evaluate the safety of the drug at different doses	20–80 healthy volunteers (except for anticancer drugs)
2.	Determine if the drug is active and tolerable in a small group of patients	100–300 voluntary patients
3.	See if the compound is more effective than standard therapy in a broader group of patients and in most hospitals	1,000–10,000 voluntary patients
Approval	The European Medicines Agency (EMA) assesses the results of the clinical study and approves or refuses the marketing of the drug. In the US the task is entrusted to the Food and Drug Administration (FDA)	After approval, a national agency establishes whether the new drug may be reimbursed by the relevant national health service.

In the case of potential antitumour agents, both groups consist of patients, because the potential adverse effects of an antitumour agent tend to pose too high a risk to permit its evaluation in healthy volunteers. When a potential novel antitumour treatment is evaluated against a cancer which does not respond to any existing therapy, there are two sides of a balance sheet to be considered by the trial oncologist. On the one side, there is the willingness to offer a therapeutic opportunity to patients for whom a reasonable alternative does not exist. However, on the other side, there is the worry of the occurrence of potentially severe toxic effects. Therefore the trial always starts with very small groups of patients, often just three people, who receive minimal doses of the agent, doses expected not to elicit toxicity.

For safety reasons each new drug which enters into phase one of a clinical evaluation has undergone a mutagenicity test in the laboratory, which is an experiment which demonstrates whether it can cause DNA modifications. The most commonly applied test is one invented during the 1960s by Bruce Ames, an American bacteriologist. The general principle of the Ames test is as follows: the test compound which might cause a DNA mutation is added to a dish containing bacteria which are incapable of growing in the absence of a particular substance, for example, a sugar. If the test compound causes one or more mutations, some of the bacteria reacquire the ability to grow even without the sugar, because their DNA has been modified by the compound. So, in this test colonies of growing bacteria indicate a "positive" result, meaning that the test compound to which the bacteria were exposed may cause mutations, i.e. it may be mutagenic (Fig. 2).

Some compounds are not mutagenic *per se* but generate mutagenic metabolites during their passage through the liver, the organ with remarkable substance-transforming ability. Therefore, before being submitted to the Ames test, a drug is often incubated with rat liver homogenate containing the enzymes capable of transforming the drug into its metabolites.

Only if the result of the Ames test or a similar test is negative, does the compound proceed to testing in a phase one clinical trial.

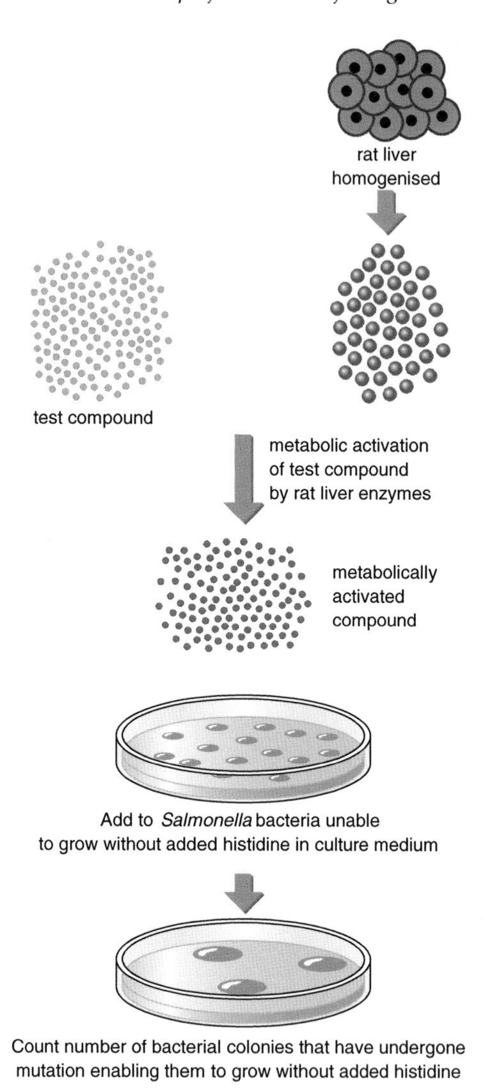

Figure 2. The Ames test to evaluate the mutagenicity of a compound (adapted from R. Weinberg, *The Biology of Cancer*, Garland Science).

This rule is strictly followed in the case of all new drugs designed to treat diseases of low or medium severity or diseases in children or pregnant women. For very severe or incurable diseases which lack efficacious treatment modes, it can be acceptable that a potential new drug may possess major side effects. There are for example antitumour drugs, which whilst undoubtedly efficacious, possess mutagenic properties. In these cases, the justification of their use in humans is that their limited tolerability due to toxicity is preferable to certain death.

The initial dose is small

For reasons of safety, the first dose of a new experimental drug administered to a human being is always very low, normally a fraction of the dose which elicited toxicity in animals, extrapolated in proportion to the weight of a human.

Such a low dose is likely to be pharmacologically inactive. The risk of lack of effect of this low dose is that it might suggest the drug is inactive. Therefore, if no severe adverse effects manifest themselves after the first dose, the dose is escalated in gradual fashion following a mathematical scheme inspired by the Fibonacci sequence. In practice, the successive dose increments are increasingly less as the study progresses. For example, if the first dose is 5 (for example, milligram or microgram), the second one might be 10, the third 12.5, the fourth 13.75 and so on.

For each dose, the concentration of the drug is measured in the blood and urine to find out about its absorption, bodily distribution and elimination.

Should any of the volunteers experience severe toxic effects, the dose which has provoked the problem is called the *maximum tolerated dose*. Then the dose is decreased by one increment, and only if this decreased dose is then tolerated, is it established as the so-called *recommended dose*. This dose remains the recommended dose, provided that it is really tolerated even after several treatment cycles and not only after the first one.

If severe toxic effects are observed, the trial is closed for ethical reasons irrespective of the phase of evaluation which the drug has reached. The medical doctors who guide the study may well be emotionally wrapped up in the experiment, and they may even be unaware of this emotional involvement. If they consider the experimental drug as their "child", they may find it impossible to abandon the trial even given the worst case scenario, that is, if severe toxicity occurs. To avoid such complications, it is important that difficult but inevitable decisions concerning the continuation of a trial are taken by an independent group of experts, who also conduct the trial analysis.

A phase one trial is not necessarily the last resort

We pointed out that the first phase of the clinical evaluation of a new drug serves to evaluate if the drug is tolerated, but it may also give some initial indications as to its efficacy, which is especially important in view of the expectation of the participating volunteers.

Sometimes, it is therefore possible to establish another essential parameter at this stage of clinical evaluation: the so-called *minimum effective dose*, the smallest amount of the drug which, when administered, elicits a therapeutic effect. For therapeutic hormones and biologicals, which have a precise target, the establishment of this dose level tends to be somewhat easier than for drugs of small molecular weight lacking a well-defined target.

In some cases, it is possible to construct a preliminary set of pharmacokinetic, pharmacodynamic and toxicity curves at the end of the phase one trial. These curves, which show graphically (Fig. 3) how some parameters vary with the increase in dose of experimental drug, are by no means always linear. Doubling the dose does not necessarily cause a doubling of the effect. For example, above a certain dose, the capacity of the organism to metabolise and thus inactivate the drug may be saturated, so that even a small increase in dose may cause a marked increase in toxicity. The same argument holds true with respect to the therapeutic effect. When the amount of drug administered into the organism saturates the number of available pharmacological target molecules with which it interacts to exert its activity, an increase in dose over this limit fails to add any benefit. This can be seen for imatinib (gleevec) in the graph below.

Evaluation of the pharmacokinetics, pharmacodynamics and toxicity information gathered together helps define the so-called

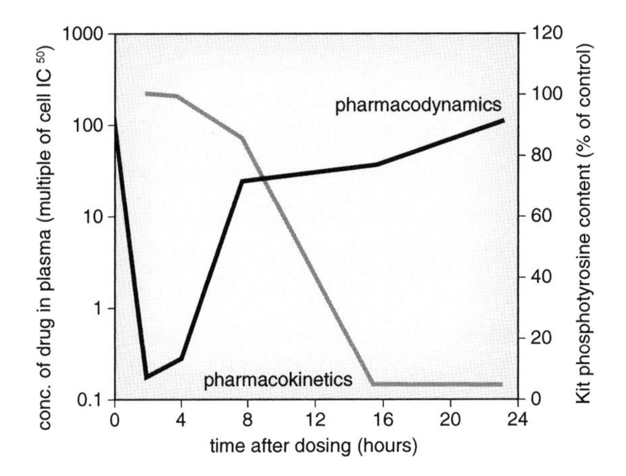

Figure 3. The pharmacokinetics and pharmacodynamics of imatinib (adapted from R. Weinberg, *The Biology of Cancer*, Garland Science).

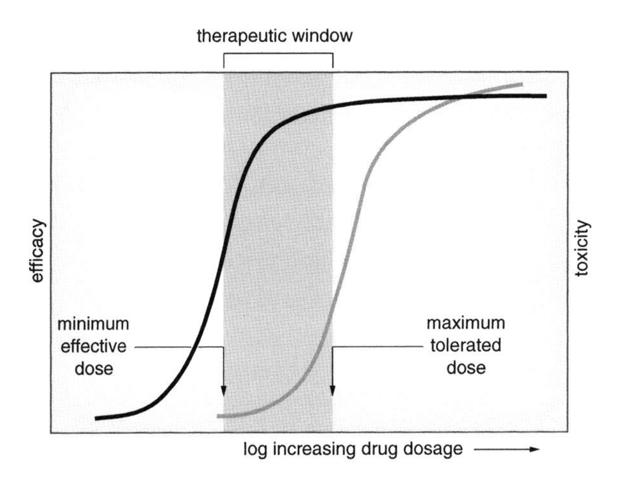

Figure 4. The therapeutic window of a drug (adapted from R. Weinberg, *The Biology of Cancer*, Garland Science).

therapeutic window, the gap between the minimum effective and maximum tolerated doses. In other words, the therapeutic window (Fig. 4) delineates doses of a drug which generate pharmacologically active concentrations in the organism without any, or with only moderate and acceptable, side effects.

Towards phase two: Drugs which just make it

In 1973, John Cleland was a student of veterinary medicine at the University of Indiana, when a rapidly growing nodule was identified in one of his testicles. After emergency surgery, the treating medics diagnosed a testicular tumour which had already metastasised to the lungs and lymph nodes, and he was given a survival probability of 5%.

Lawrence Einhorn, who at the time was a young oncologist (Fig. 5) who treated Cleland, wanted to try everything in his power to save his patient. But the weapons at his disposal were very limited. The only feasible therapy option was a cocktail of certain chemotherapeutic drugs based on the results of previous clinical studies. He knew he could do little against Cleland's tumour.

Cleland had already lost a third of his weight and could not stand on his feet because of the awful side effects of the treatment. Einhorn decided to try a new combination including cis-platinum, a drug which had shown some beneficial effect against testicular tumours — this is one of the drugs discovered by chance, as you may recall, which we talked about in Chapter 1.

Cleland received cis-platinum, and it did stop the growth of his tumour cells as he was literally on death's door. After only ten days

Figure 5. Lawrence Einhorn

of treatment, Cleland went into complete remission, and the metastases disappeared as by magic.

Yet, cis-platinum was very difficult to tolerate. It made patients vomit 12 times a day, on average, and it caused problems to the kidneys. At the end of the early phase one trials against several tumour types, more or less all oncologists involved with these trials were inclined to ban it to the shelf marked "unsuccessful experimental antitumour drugs", because of grave toxicity and only scarce therapeutic efficacy.

Einhorn saved cis-platinum from therapeutic oblivion. In 1975, he presented results obtained in 20 other patients suffering from the same type of disease, testicular cancer, which had been, up to that moment, a truly brutal disease. Using this new drug, he changed everything.

Nowadays, more than 30 years later, cis-platinum is the standard therapy for testicular tumours, 95% of which it cures. It is also used against other tumour types, albeit with less impressive results. The kidney toxicity has been reduced by simply making patients drink a lot and by administration of liquids before and during therapy, whilst the intolerable vomiting it can unleash is nowadays controlled by drugs against nausea.

An uncertain therapeutic effect on the one side and the high likelihood of grave side effects on the other render selection of a new anticancer drug extremely difficult. The lessons learnt from cis-platinum are therefore stern reminders to approach such trials with optimism, trust and humility. Optimism that it is possible to discern at least some positive results; trust in finding suitable antidotes against side effects, which are not available against death; and humility to imagine that even disappointing results may conceal novel treatment opportunities for some patients, even if they can often not be seen clearly.

Phase two

The second phase of the clinical evaluation of a new drug has two aims: To learn whether the drug has some beneficial effects on the disease, as hypothesised on the basis of the preceding preclinical research, and to continue monitoring its safety.

In practice, a phase two trial proceeds as follows: Initially a small number of patients is recruited, for example 20, just enough to serve as examples. Whilst healthy volunteers can be involved in a phase one study, they cannot in phase two, here the volunteers are always patients. All recruited patients receive the same recommended dose of the experimental drug or the placebo. If at least one or two of the patients in the group receiving the experimental drug exhibit a response, further patient volunteers are recruited into the study. Thus the study is extended bit-by-bit, thus avoiding exposure right from the start of a considerable number of individuals to any potential toxic effects.

The types of drug activity which can be observed in a phase two trial are attainable and measurable rapid responses. For example, in a trial of an antitumour drug, the size of the tumour can be examined, establishing whether it shrinks, or at least fails to enlarge, in a significant percentage of patients. Obviously the most important insight would be not tumour reduction *per se*, but whether the drug is more active than previous therapies in controlling the disease and prolonging the patient's life. To study these issues, a longer and more expensive clinical trial is required, the so-called phase three trial.

Phase three

The point at which the "cancel" command, deciding on continuation or discontinuation of the clinical experiment has to be seriously considered, is at the end of the phase two trial. At this stage the question is asked: does the drug really deserve to be subjected to the ultimate clinical test?

In the subsequent phase three trial, the clinical evaluation is extended to include a very large number of patients, usually hundreds or thousands. The aim is to explore if the results obtained up to then can be confirmed, or indeed dismissed, in a large population of patients.

Each patient recruited into a phase three clinical trial is being followed up for years. Sticking with the antitumour drug example used above, whilst in phase two it would be sufficient to observe a measurable effect, such as tumour regression, this is not sufficient in phase three. Here one wishes to observe whether the drug increases the patients' survival in comparison to drugs currently in clinical use against the disease under study. Often, a phase three study of a new agent is conducted in parallel in several medical centres, because this is the best way of finding out if previously obtained results can be confirmed in different environments with different medical doctors, health workers and patients.

The phase three trial is very expensive, because the total health service expenses for all recruited patient have to be paid for by the experimenters. In comparison, phase one and two trials are small fry. For this reason a pharmaceutical company reflects exceedingly carefully before it initiates a phase three trial.

As we are talking about money, it is worth adding that only phase one trials can be supported by a public or non-profit research organisation. On grounds of both costs and organisational requirements, investment by and experience of a pharmaceutical company are usually required to perform studies from phase two upwards. The costs for a phase three clinical study can easily exceed several tens of million pounds.

Endpoints

The objectives of phase three trials are defined in terms of experimental drug efficacy indicators, called *endpoints*. Studies which have the measure of *survival* as an endpoint are the most rigorous, because the time remaining in a patient's life is a clear-cut, irrefutable and unambiguous piece of information.

Most clinical studies establish endpoints of a somewhat less definitive nature, as for example *disease-free survival*. What does that mean? It means that the drug retards the progression of the disease, whilst it may not improve survival when compared to another drug in current use.

More ambiguous are the so-called *surrogate* or *proxy* endpoints which measure, for example, a biochemical parameter in the blood theoretically linked to survival or to disease-free survival. A classic example of this type of endpoint is the measure of cholesterol in the blood in clinical studies of statins, which are drugs designed to reduce the impact of cardiovascular diseases by limiting the accumulation of fat in the arteries. In the study of a statin the most rigorous endpoint would be the patients' survival. But the problem is that, to follow a large group of patients for several years before the true endpoint is reached in every single one of them would be extremely expensive. Reduction of cholesterol levels is a measure which can be obtained at much lower costs and within a much shorter time frame. However, this surrogate endpoint is not always associated with more definitive endpoints in these patients. In many cases cholesterol levels can be reduced without subsequent increase in survival. In other words, the compound under test is pharmacologically active as it reduces cholesterol levels, but not therapeutically active, as it fails to alter survival.

So we must appreciate that biochemical parameters are undoubtedly valuable, so that in each clinical study many such parameters should be, and are, appropriately measured. But if used as surrogate endpoints, they are only a crude and sometimes even misleading measure of the efficacy of a drug.

Not Inferior Drugs

Each year hundreds of new drugs are being approved and introduced into the health market, but that does not mean that they are all really novel. Clinical studies of new drugs can have as endpoint not their superiority but their *non-inferiority* when compared to a standard treatment used in the particular disease under study. Some experimental drug molecules submitted to clinical experimentation are only slightly modified versions of old drugs, the patents of which are about to run out. In this case, a costly clinical study is conducted, inevitably exposing volunteers to potential toxic effects, with the objective not to improve on the efficacy of an existing drug, but to obtain a new drug molecule which is often only what insiders would sarcastically call a "me-too" analogue. This approach strikes many as rather dodgy.

Nevertheless, one could also argue that, especially in the case of chronic diseases, the objective of the development of a new drug may not only be to attain better efficacy, but also to achieve reduced side effects, or an improvement in mode of administration. Therefore new drugs can constitute a considerable medical advance if they are, even if not superior to existing medications, less cumbersome to handle or easier to take, for example, only once instead of three times a day.

"One Can't Predict the Weather..."

Clinical trials of many novel drugs, the development of which have been accompanied by enormous expectations, demonstrate only very modest efficacy improvements when compared to drugs in current use. In some cases survival is increased by six months, in others by three and in yet others by one week only.

"One can't predict the weather more than a few days in advance" — this quote by the theoretical physicist Stephen Hawkins, used in the context of black holes, illustrates how insignificant such progress can appear, especially in the light of the hype which sometimes surrounds trials of new drugs and the staggering prices often charged for them by pharmaceutical companies. A single therapy cycle with a new agent can cost amounts of money bearing five zeros, whilst the costs of traditional drugs seldom exceed numbers with two or at the most three zeros.

In practice, many novel drugs work reasonably well in a minority of patients, yet are without activity in all the other patients who receive them. Patients who respond are those who possess the specific receptors or other biomolecular characteristics necessary for the drug to exert its action. Therefore clinical trial survival data can represent a kind of average of two different situations: Increased survival in the few patients with the right biochemical characteristics for the drug to work, and survival equal to that offered by traditional drugs in the many who do not have these characteristics.

A solution to this trial dilemma is to test such drugs in small groups of patients selected on the basis of presence of the molecular signature required for drug activity. An example is herceptin which we have already mentioned in Chapter 2. This anti-breast cancer drug works in only those women whose tumours overexpress a receptor called Her2. When used in women with this biochemical characteristic, the drug is active, but the measure of efficacy in trials is diluted if herceptin is administered in an indiscriminate fashion to all women with tumours both positive and negative with respect to Her2 expression. In other words, efficacy of novel drugs can be observed when patients are *stratified*, that is subdivided into groups with defined molecular characteristics.

A problem of conducting such a study in small numbers of patients is that statistical validity is difficult to reach. Nevertheless, this obstacle can be overcome, at least partially, as we shall see in the next chapter.

Chapter 5
Only Probabilities, Never Certainties

A drug administered at an identical dose to different individuals can be recovered from the blood of these individuals at concentrations which can vary 10- to 20-fold.

If you have read thus far you know already that this amazing difference is caused by metabolism, this means the process in each of us which transforms the agents which we ingest at very variable rates. Metabolism is in turn influenced by the diet, by other drugs we may take, by body weight and so on. For example, grape fruit juice is known to interact with the metabolism of some drugs, and excessive body fat can sequester drugs so that they can be stored there for a long time. In addition, there are differences in our genetic make-up which can account for differences in metabolism.

Apart from the known suspects, there are very many unknown factors capable of interfering with the activity of a drug and of confounding the outcome of clinical studies.

One of the major issues which we need to take into account when we interpret the results of a clinical experiment, and even more so, when we consider the consequences of taking a drug, is that each person is different from the next.

For example, when one reads that a drug works in 72% of cases suffering from a particular disease, it means that on the basis of the results of a clinical study, given that a large enough sample of patients was enrolled, the drug was active on average in 7 out of 10 persons. The percentages given to describe clinical outcomes are always a *probability* and never a *certainty*. Were the effects observed in these 7 out of 10 individuals all the same, or did they all react somewhat differently from each other? And in my body, how would this drug behave?

The response to a drug, assuming there is one, varies tremendously from person-to-person. In my case it may have a complete effect, work to some extent, or not work at all. Or it could be, that my symptoms get better irrespective of the fact that I took the drug, because many illnesses resolve without any intervention, just think of the flu. Or the placebo effect may have played a role, as discussed in Chapter 4.

The medical doctor and the pharmacologist, who study new drugs, always face situations of this sort, in which it is really difficult to attribute only those effects to the drug which it has actually caused and not those generated by other confusing factors.

So, for those in the business of testing new drugs, mistakes, errors, variability and surprises are run-of-the-mill episodes. In this chapter, we shall learn about how statistics can help the pharmacologist to avoid errors and to understand any oddity within the results of a study. If you dislike numbers, do not worry. We talk only about principles and do not describe mathematical formula or details, and we discuss these principles only in "homoeopathic" doses, as much as is necessary to comprehend what it is all about.

Measurements are Never without Errors

We suggest a simple experiment to you. Take a fever thermometer and measure several times your temperature. Three or four measurements repeated one after the other are enough to note small differences in the temperature which you have measured. One would expect that the measurements do not differ much from 36.5 °C if you feel well. But it is unlikely that they are identical. Within certain limits your body continues to generate slightly different temperatures. Also, you may have introduced variability into the measurement which may have influenced the results, for example, by the way in which you inserted the thermometer, or by the time period you spent on the measurement, or also by the care with which you took the temperature.

Scientists who evaluate new drugs are confronted constantly with variabilities of this type. For each result observed, they need to ask themselves: Is what I have measured a real effect of the drug, or is it a mistake, or have I in some way put my foot in it?

Is the Study Sufficiently Powered?

A way to reduce the weight of error and variability is to perform very many measurements. If we test our new drug in thousands of patients rather than in tens or hundreds, we can be a little more certain that the results which we obtain are significant and not erroneous or the effect of some rare variant.

To recruit a large number of volunteers is, however, not an easy enterprise. According to a study conducted at the Fred Hutchinson Research Centre in Seattle, only 3% of patients qualifying for inclusion accept taking part in a study, and about 40% of experiments fail because of the impossibility to recruit sufficient numbers of patients.

Therefore the question is: what is the minimum number of patients who have to be recruited into a study so that their number is sufficient to confirm or reject the hypothesis that our drug is active? The answer depends according to whether we wish to demonstrate a large effect or a small one.

Here is an example: We want to prevent a very wide-spread disease, such as malaria, with a vaccine, which has shown strong activity in phase two, in around 90% of the patients. Such a big effect should easily be apparent in phase three even if a relatively small number of individuals is recruited.

If instead the vaccine we are evaluating has shown efficacy in only 10% of cases in phase two, we need a very large number of individuals to prove in phase three that our hypothesis is correct.

The power of the study is the numerical capacity of a test to refute the so-called *null hypothesis* that is the possibility that our drug *fails* to elicit the activity which we expect. If the power analysis is ignored and the number of patients enrolled in a study is too low, the study is called *underpowered,* in simple terms it means the effect may be there, but one does not see it. Unfortunately many clinical studies fail and many new drugs do not reach the stage of proper development and the patient's bedside because the statistical analysis of the phase three study is deficient.

You may ask yourselves, why drugs with only a low degree of activity, let us say at the level of 10%, may be of interest. Malaria is a good example. The World Health Organisation (WHO) estimated for 2012, 207 million cases and 627,000 deaths of the disease. If a vaccine or a drug were able to reduce the number of those afflicted with this disease by even only 10%, 20 million people would benefit, more than the inhabitants of Shanghai!

If all these probabilities remind you of placing bets on horses, you are not entirely wrong: Each hypothesis is partly a gamble, because nobody possesses the magical crystal ball, and unknown factors are numerous. But a crucial difference between a pharmacological hypothesis and a bet placed on a horse is that the former has a rational basis built on solid information about the drug and the disease accrued in years and years of laboratory research before the trial in humans began. The latter bet has no rational basis!

Is the Result a False Negative?

Imagine that we evaluate a new drug in phase two, and after having been injected into a certain number of patients, nobody gets better. Can we therefore conclude that the drug is inactive? No, the only certain thing which we can say is that we are not able to measure any effect. We may deal with an effect which is too small to be measured and thus a *false negative.*

The most important antidote against this kind of error is to make sure the study has adequate power, which means it involves a sufficiently large population of patients. But there are other explanations, which have to do with *insufficiency*. The dose of the new drug may have been too low, the duration of the therapy too short, or the period of observation of activity too brief.

Therefore it is important to understand in this situation whether the drug is genuinely inactive, or some parameter, either the number of patients, dose, time of administration or duration of observation, have been incorrect or inappropriately assessed.

Is the Result a False Positive?

This is the opposite case. After injection of our new drug into a certain number of patients we observe that they get better. How do we go about making sure that the effect is genuinely due to the drug and not to a placebo effect, to a spontaneous improvement in the disease or to the action of another drug that has been administered at about the same time?

Human pathology is very variable. One has to realise that the number of cases in a clinical study is inevitably very limited when compared to that of all cases occurring in the real world. Therefore results of a clinical study can be ephemeral and difficult to interpret.

So we may be faced with a *false positive* result. If we judge the drug as being active and accept this result without asking further questions, we would reject the null hypothesis, which implies that the drug is inactive, even if this is correct.

Without getting embroiled in a complicated description of statistics, it is probably sufficient to realise that there are procedures and relevant software which permit assessment of the influence of confounding factors thus reducing to a minimum errors which these could cause.

Can We Exclude Chance?

One issue in a clinical study is especially difficult to control: The results should reflect a true difference between the two arms of the study, and any difference should not be the consequence of chance.

Here is an example: In a clinical study of a new potential antidiabetic drug, we find that the activity of the new drug occurs in 18.5% of patients, whilst the traditional treatment is active in 8.3%. How can we make sure that the difference is due to a real effect of the new drug and not to chance?

Let us conduct a thought experiment with a coin. We toss it in the air 30 times, noting the number of heads or tails coming top each time. And then we repeat the experiment many times with the same 30 tosses. In most cases the numbers of heads or tails coming up top are more or less close to the numbers expected on the basis of pure chance, i.e. 15 heads and 15 tails. We might observe for example 13 and 17, 16 and 14, 17 and 13 and so on. But there are sometimes results which deviate from the norm, let us say 7 heads and 23 tails, or 21 and 9, and so on. The issue is how frequent are such "strange" results compared to those dictated by chance? This depends on where we place the bar between "strange" and "not strange". Conventionally one considers values as deviations from the norm when they occur in fewer than 5% of tosses.

Let us return to our new antidiabetic drug. We have to reason more or less on similar lines. If the results are not the consequence of chance, the number of measurements which deviate from the norm, i.e. is from the majority of the observed values, must be less than 5%. Only in this case can we say that our results are *significant* at the 95% probability level. This is the same as saying that there is a 5% probability that we made a mistake, and nobody can claim that our results are all among these 5%!

Medical statistics involve in reality much more complicated calculations to establish that the results are significant, but the empirical reasoning given here gives you the basic idea of how one goes about dealing with chance.

Cause and Effect or Correlation?

When we observe an effect in a clinical study, it is easy to ascribe the effect to the new drug. Yet the effect might have been a fortuitous coincidence, or it could have been provoked by a so-called *confounding factor*, a variable completely unrelated to any drug effect and thus confounding the results.

The link between smoking and lung cancer may serve as an example. It is not pertinent to any drug, but might help to clarify the concept.

After the use of cigarettes had spread among the soldiers and veterans of the First World War, lung cancer epidemics — previously a rare disease — started to emerge wherever people smoked. There were however novelties other than cigarettes which gained popularity at that time, for example, women's tights. Tongue in cheek, a medical doctor who might have been unconvinced by the hypothesis that smoking causes lung cancer could have said that also the use of women's tights had increased in parallel with lung cancer incidence. In other words, there would have been a *statistical correlation* between the application of women's tights and lung cancer, meaning numbers of incidences increased in parallel, yet there was neither a *biological correlation* nor a plausible link between cause and effect.

Statisticians have established a series of questions, the answers to which have to be positive if a correlation can be taken to mirror a genuine link between cause and effect.

(1) *Is the correlation strong?* Yes, the risk of cancer among smokers is at least 5–10 higher than among non-smokers. In contrast, the risk is low among either those using or those not using women's tights.
(2) *Can the correlation be repeated in a systematic fashion?* Yes, the more one smokes the higher is the incidence of cancer in any environment and any population. The same does not apply to the use of women's tights.
(3) *Is the correlation specific*? Yes, tobacco consumption is strongly associated with lung cancer which is a well-defined disease in a specific organ, into which the smoke easily penetrates. Women's tights are found on legs and, if used appropriately, nowhere near the lungs.
(4) *Is the correlation linked to length of time?* Yes, the greater the number of years during which one smoked, the higher the risk

of developing lung cancer. Such a link does not exist for the length of time of using women's tights.

(5) *Is the effect quantitative?* The higher the number of cigarettes smoked, the higher the risk of succumbing to the disease. The risk of falling ill is unlikely to vary with the number of women's tights used.

(6) *Is the correlation plausible?* Yes, because one can hypothesise a plausible mechanism, that inhaled carcinogens contained in tobacco smoke can generate malignant changes in the lungs. It is near impossible to formulate an analogous hypothesis with respect to women's tights.

(7) *Is the correlation consistent with experimental results?* Yes, the statistical data are in accordance with laboratory results obtained in studies of effects of cigarette smoke on cells and animals. It is unknown whether, and unlikely that, anybody has tested the effect of women's tights in the laboratory.

(8) *Does the correlation occur similarly in analogous situations?* Yes, cigarette smoke is associated with cancer of the lips, throat, tongue and oesophagus, apart from lung cancer, all similar diseases occurring in regions of the organism which come into contact with cigarette smoke. There are no analogous effects in the case of women's tights

Now, we have tried to illustrate the link between cause and effect using the relationship between cigarette smoke and lung cancer. We must ask the same questions when we wish to establish whether or not in a clinical trial an effect observed can be attributed unambiguously to the new drug under test.

Rare Diseases

A disease is defined as rare if it occurs in not more than in five among 10,000 persons. Such a small number of patients can render conducting new drug trials with adequate power and obtaining reliable results difficult. As an aside — it is likewise challenging to convince a pharmaceutical company to produce and market a drug against a disease which occurs in only small numbers of patients.

These are serious problems, but a statistical technique called *Bayesian analysis* may offer a solution. Whilst this type of analysis is very complicated, and detailed explanation is beyond the scope of this book, it may suffice to say that Bayesian analyses exploit acquired evidence which gradually emerges from the trial and continually modifies the strength of assumptions and hypotheses.

Solutions facilitating studies of rare diseases are important for two reasons. The first is that diseases being rare does not mean that people afflicted with them are that uncommon. There are at least 7,000–8,000 known rare diseases, and many more are still to be discovered. Adding up the numbers of patients suffering from all known rare diseases yields a large number!

The second reason is that we understand more and more about molecular causes of rare diseases. Insights into causes can help us find out, for example, whether disease X, which is not rare, consists in reality of two diseases Y and Z, which on their own are rare. Such knowledge may eventually help us to treat patients with diseases Y and Z much more accurately and effectively than if they were grouped together as disease X.

Extrapolation can be Dangerous!

Having observed some effect in a small sample of patients, we may speculate that the same effect may be reproduced in a much larger number of patients or in another study under similar — even if not identical — conditions. Statisticians call this conjecture *extrapolation*, and it really generates anything but certainties. Only a follow-up trial conducted with a much larger number of patients or under different conditions can tell us whether the extrapolation is accurate or not.

This chapter gives you some idea of the immense caution which needs to be exercised in clinical studies when numbers are handled. Both at the beginning, where we show how correct numbers support the power of a study, and at the end, where we describe that reading numbers with a critical and sceptical mind helps to interpret results correctly. So, as one may say, numbers given or read erroneously can severely damage health, but the argument can be turned round: well-planned and carefully considered numbers can ultimately save human lives.

Chapter 6
Approval or Rejection

In 2013, a large American pharmaceutical company submitted an application for approval of a new sleeping pill to the Food and Drug Administration (FDA), the US authority which clears the path to the use and commercialisation of new drugs. The application, submitted in electronic format, took as much as 41 gigabytes digital storage space. For comparison, all Wikipedia entries in 2013 together took up 9.2 gigabytes. The dossier for the new sleeping pill was not exceptional. The documentation describing the efficacy and safety of every new drug seeking approval needs to be more than encyclopaedic.

The process which ultimately leads to approval starts at the end of the phase three trial. The pharmaceutical company which wishes to commercialise an experimental drug successfully in a specific country needs to submit a dossier to the regulatory authority of this country describing in the most minute detail not only whether the drug is safe and efficacious and to which extent it is pure, but also in which dosage form, at which dose level and for which indication it should be prescribed.

In this chapter, we shall see how the approval of a new drug comes about, and why the documentation has to be so substantial. In order to understand the logic behind the approval process, we need to take a step back in time and find out how the control authorities which evaluate and authorise drugs in the world came about.

Who Does Approve New Drugs?

At the beginning of the 20th century labels of many medicines contained information claiming they would cure more or less anything, from cancer to infertility, from tuberculosis to epilepsy with any conceivable "female indisposition" thrown in for good measure. Seen through our eyes the promises were clearly absurd. Furthermore, there was no mention of either any side effects or composition of the medicine which was an industrial secret, just like the recipe for Coca Cola.

In 1905, the American journalist Samuel Hopkins Adams, published a series of 11 articles in *Collier's Weekly*, a contemporary weekly magazine, in which he unmasked many of the declarations made on the labels and revealed that many ingredients contained in medicines were dangerous.

Adams' articles provoked such a scandal that already a year later, in 1906, Congress approved the Pure Food and Drug Act, a law which stipulated the purity of foodstuffs and drugs, prohibited the sale of adulterated foods and medicines with misleading labels and eliminated the possibility to keep the composition of a medicine secret.

Only a short time later, the FDA was born, a government agency which had the responsibility to correct the deplorable situation which existed. The FDA as it was immediately called had to investigate all commercially available food products and pharmaceuticals and to establish which of these were beneficial and thus suitably available and then to eliminate the rest.

In 1938, another scandal led to another new law, the Federal Food Drug and Cosmetic Act. Sulfanilamide, an antibacterial agent formulated in powder or pill form, was shown to be effective against streptococci. Doctors wanted to prescribe it in the form of a syrup, and after many attempts Harold Cole Watkins, a chemist with the company Massengill & Co, who produced the drug, succeeded in dissolving the drug in a solution containing diethyleneglycol. The syrup with raspberry taste which was called sulphanilamide elixir, was distributed to all pharmacies in the whole of the US in 1937. But after only a few weeks, some deaths were reported in people who had taken the medicine. The syrup was immediately recalled, but 105 of the 353 people who had received it died of renal insufficiency. The safety of this medicine had not been investigated, because proof of safety was not requested by law up to that time. Also the name of the drug was incorrect, an "elixir" is supposed to contain alcohol, but in the fateful syrup there was not a trace of alcohol.

The disaster of the sulphanilamide elixir had dire consequences. The chemist who had formulated the syrup committed suicide. The company was fined to pay 500,000 dollars, the equivalent of about 8 million of today's dollars, to the families of the poor patients who had died of the medicine.

Even though one would of course prefer laws to be implemented before, rather than after, any scandal, the 105 lives which the sulfanilamide syrup claimed had a positive and enduring legacy. The new law of 1938, empowered the FDA to demand from the producers of new drugs proof of safety and efficacy as a condition to obtain authorisation of commercialisation.

The rules on safety, efficacy and purity of drugs established by the FDA have for the last century been the reference point for all major drug regulatory authorities in the world. Most of these agencies, among them the European Medicines Agency (EMA) created in 1995, have shaped their organisational structure and rules on the American model, because the FDA was the trailblazer in defining the control and regulation of drugs.

Nowadays, the EMA, which is currently located in London, is the only agency which can authorise the use of a new drugs in the countries of the European Union (EU), after conducting an evaluation process which must not exceed 270 days. The drugs approved by the EMA can enter the market in all EU countries, and no country can oppose this decision.

After EMA approval the agencies in individual EU countries come into play. In the UK this is the Medicines and Healthcare Products Regulatory Agency (MHRA), in Germany the Bundesinstitut für Arzneimittel und Medizinprodukte (BfArM), in France the Agence Nationale de Securite du Medicament et des Produits de Sante (ANSM), in China the China Food and Drug Administration (CFDA) and in Japan the Pharmaceuticals and Medical Devices Agency, to name but a few such bodies. They have to approve commercialisation in their countries of each new drug endorsed by the EMA, and recommend whether their prescription should be funded by the respective national health services.

Whilst the prime task of the national regulatory authority is only a bureaucratic formality — as we said, a EU country cannot but grant commercialisation of a new drug approved by the EMA — the second task is a rather delicate one, because economical calculations may well be inconsistent with ethical considerations.

This task entails measuring the relationship between the efficacy of a new drug and its costs. If, for example, a drug is only weakly active and very expensive, the drug regulatory body may decide to refuse funding of treatment with this drug. Resources of national health services are limited and priorities need to be set.

Thinking about the community at large, the basic objective here is a dual one: offering as many efficacious remedies as possible to citizens who need them, and guaranteeing at the same time the economic viability of a universally accessible health system, such as those which exist in most European countries. The perspective changes somewhat, when one has single individuals in mind. Should a person suffering from a rare disease have to pay the full costs of a drug because it is considered low priority, whilst a patient afflicted with a common disease has the right to obtain a drug free of charge because it is judged very important? And then, what is the value of drug-induced added survival in perhaps only a few humans lasting perhaps only a single day? Perhaps health economists have answers to these questions. We do not.

Stories of Rejections

In the FDA building, outside the offices where the representatives of pharmaceutical companies meet agency staff to discuss potential approval of a new drug, samples of toxic compounds which the

FDA is proud of having rejected are prominently on show. These drugs have made history because of their horrible toxic effects, and they are a most poignant reminder for all those present asking for the approval of a new drug of the fact that safety is an even more important consideration than efficacy.

Among the drugs which stand out in this frightening collection is thalidomide. It was introduced into the European market in 1956, first as a flu remedy then as a tranquilliser and antisickness drug in pregnant women. Thalidomide was approved in Great Britain in 1958 and already in 1961 it was used in at least twenty European and African countries.

In 1960, the pharmaceutical company which produced thalidomide submitted a request to the FDA to consider approval of thalidomide for its commercialisation in the US. In the clinical study demanded by the FDA, 2.5 million tablets had been administered to about 20,000 patients. In the course of this trial, the FDA received indications that 17 children of patients were born with severe limb malformations. Frances Kelsey, the pharmacologist at the FDA who had evaluated the request for approval, recommended that thalidomide should not be approved. For that courageous stand, she was later honoured by President Kennedy.

In other countries in which thalidomide was widely used, more than 10,000 children with deformed arms or legs were born before the drug was withdrawn from the market in 1961.

The utterly disgraceful thalidomide affair led to very useful new regulation. Drug regulatory authorities turned their attention to *teratogenesis*, the ability of certain chemicals to interfere with embryo development. From the 1960s onwards it has been stipulated by law that new drugs must be evaluated in terms of their potential effects on foetal development involving experiments in animal species whose gestation is sufficiently similar to that in humans.

In the 1990s, thalidomide experienced an unexpected renaissance because of its beneficial effects against certain tumour types, in particular multiple myeloma. This is another example that a poison may also be a remedy, depending on dose, disease and the individual who receives the remedy.

Which Criteria are Important for Drug Approval?

Quality, safety and efficacy are the three key words which help secure approval of a new drug. Let us start with quality, which we have not mentioned before. A drug is of good quality if its composition is always the same, if it is stable for a defined time period thus allowing setting an expiry date, and if it can be absorbed by the organism.

We have already dealt with safety and efficacy extensively, and you should have a good idea of what these terms mean. Nevertheless, let us briefly recapture. A drug is safe if its side effects are known and acceptable. A drug is efficacious if it is capable of exerting a beneficial therapeutic effect against a defined health problem.

Not all drugs can be judged and thus approved or rejected using the same criteria, and this is for good reason. For example, side effects of a drug developed for a minor disease are not acceptable, as these might cause more harm than the existing disease. Drugs for children or women of fertile age must be extremely tolerable because it is unacceptable to cause toxicity in youngsters who have many years of life before them or in women bearing cells in their body which are the springboard of a new human being.

The graver the condition to be treated, the greater the acceptable level of toxic side effects, because with the alternative being death, an unpleasant side effect is surely the lesser of two evils.

Good Practices

You may recall the size of the dossier which needs to be submitted to regulatory agencies which approve new drugs. A considerable portion of this documentation deals with some practices which have become international standards endorsed by these agencies.

For example, the *good clinical practice* imposes a series of rules on how to design, conduct register and describe studies in human beings. The aim is to protect patients from poor-quality clinical experimentation, safeguarding their wellbeing and the confidentiality of information related to them. In analogy, there are the *good laboratory practices* and the *good manufacturing practices*. They lay down well-documented, verifiable and reproducible procedures, which define unambiguously the responsibilities for the conduct of experiments to establish safety and efficacy and for the production of drugs.

The principles behind these practices respond to the expectations of citizens of many societies who insist on highest safety, complete transparency and optimal confidence with respect to any products they acquire, but in particular to drugs they receive.

The good practices have made clinical experimentation very long-winded, complicated and costly. But above all they have opened up some ethical dilemmas. Let us take, for example, doxorubicin, arguably one of the most efficacious antitumour agents ever developed. It has saved several millions of human lives and is still in common use after more than 50 years in the clinic, in spite of severe side effects on the heart which it can cause.

Many pharmacologists think that if approval was to be sought today for doxorubicin, it would unlikely comply with currently stipulated safety prerequisites. And the same very probably holds for many other drugs which were approved when the rules were less restrictive.

Now, this situation generates questions. Is a high safety threshold always advantageous for patients or have safety concerns perhaps engendered the loss of opportunities for the approval of efficacious drugs for diseases for which there are not yet any remedies? We do not have answers to these ethical rather than scientific questions. Such issues create doubts and concerns in our minds, but it is really up to society to find appropriate solutions.

Has a New Drug Got to Be Innovative?

In a typical pharmacy, there are about 14,000 medicinal products available, as mentioned at the beginning of this booklet. But not all of these products are unique. The indications for many of them are identical, many products are designed to tackle the same medical problem. Therefore the regulations in force cannot and do not request that new drugs need to be innovations and superior to old ones. Pharmaceutical industry often asks for approval of remedies which appear to be novel but are in reality a "restyled" product or a "me too" drug, which we have touched upon at the end of Chapter 4.

In order to establish superior efficacy, comparative studies need to be conducted in which one group of participant volunteers receives the new compound A, whilst the control group takes drug B, a remedy in current use. This is the only way to demonstrate whether one drug is superior to another. But such a procedure is unfortunately not required by law.

Transparency: Needs to be Improved

As said before, the dossier which needs to be submitted to obtain new drug approval is encyclopaedic. Now, how much of the information contained in this document is accessible to the general public? In the US, the FDA can allow access to new drug documentation after the motivation for the request of disclosure has been evaluated and considered justified. In contrast, in Europe, new drug dossiers are deemed secret documents, and the reasons for this are associated with the intention to protect pharmaceutical companies from their competitors.

Not surprisingly, public agencies entrusted with the regulation of drugs are often caught in the middle of several organisations with often highly conflicting interests. Firstly, there are the pharmaceutical companies and their shareholders who are hell-bent on protecting industrial secrets, profits and their employees. Then, there are the medical doctors, who want to know exactly what it is they prescribe for their patients. There are the national health services which must obtain all the available information on drugs, the costs for which they are supposed to reimburse. Last but not least, there are the patients who often exert considerable pressure on drug-regulatory agencies to get on with clinical trials of new potential drugs. Therefore the regulatory agencies' claim to totally correct and comprehensive information is undoubtedly justified and really important.

To arbitrate on such issues of high public sensitivity is not easy. In Europe, some medical scientists involved with new drugs argue that the EMA should make experimental clinical data accessible to the general public — at least those on efficacy and safety, without companies having to divulge sensitive information such as new drug production methods. These scientists advocate radical reforms including obligatory publication of the results of all authorised clinical studies which have been conducted on new drugs, also those with negative outcomes. But a new law would be required for this to happen.

An initiative called *All Trials*, led by the English epidemiologist Ben Goldacre and others spearheads such an approach, but it is deemed by many too radical to be successful. Publication of all clinical data might be a demand beyond the realms of the possible. Nevertheless granting unsparing and comprehensive transparency could save some pharmaceutical companies considerable embarrassment.

Unpublished Studies and a Treatment for Depression

Sometimes the lack of transparency can generate uncomfortable stories. Drugs are advertised for their tremendous efficacy, and then it transpires that in reality they are only slightly more active than a placebo, when access has been granted to the complete documentation allowing perusal of all studies including unpublished ones with negative results.

This actually occurred in the case of the so-called *selective serotonin reuptake inhibitors* (SSRI), which are among the most frequently prescribed antidepressant drugs in the world. The scientific name does probably not mean a lot to you, yet you might have heard of prozac, the commercial name of the most widely known of these drugs. The active principle of prozac is fluoxetine, which has been available for more than 25 years. About one citizen in ten in the Western world receives a prescription of fluoxetine or a chemically related antidepressant.

How do these drugs work? For decades the theory has been that they augment the concentration in the brain of serotonin, a neurotransmitter substance which stimulates the activity of certain neurons involved with the regulation of mood swings. Today, we know that this explanation is rather insufficient, given that another major effect, the formation of new neurons, is observed in the brain of individuals in whom these drugs are active.

How much we do or do not know about their precise mechanism of action is of course not necessarily a problem. We have said earlier in this book that lack of mechanistic knowledge applies to many other drugs, and it does most certainly apply to agents which act on the brain, arguably of all human organs the most difficult to study.

However, what is worrying is the small percentage of patients in whom SSRIs work, only 20–25% of those with the most severe forms of depression. In almost all other patients, SSRIs are more or less as efficacious as a placebo. We are talking here about millions of people who may embark on taking a SSRI to combat severe grief, job termination, divorce or another drastic life event from which people generally recover without treatment.

The pharmaceutical companies which commercialised prozac and related SSRIs knew right from the beginning about the limited efficacy of these drugs. But they kept the clinical trial data rigorously secret, as the law indeed permits.

This scandal came to light thanks to a group of English, Canadian and American psychiatrists. They had obtained access to the documents in custody of the FDA and analysed the results of all clinical trials conducted on these drugs. In 2005, they showed in a paper published in the *British Medical Journal* that differences in efficacy between these antidepressants and placebo were indeed negligible. So small, that the massive use of these drugs is truly unjustifiable. Especially so, in the light of the severe and often accumulative side effects associated with the sometimes long periods of time for which these antidepressants can be prescribed.

The large number of prescriptions for these drugs is further inflated by their prescription for indications which differ from those listed on the patient package insert. The issue behind this practice is called *off-label* drug use, and it is also pertinent to other drugs, as we shall discuss in Chapter 7. It is particularly relevant in the context of antidepressants, because depression is often mistaken for some other disease, and this should be diagnosed and treated in a different way, not by the prescription of a drug which was evaluated and approved for something else.

The Patent — Branded and Generic Drugs

In most so-called developed countries, each new drug approved for usage in medicine is protected by a patent. This is a document which establishes who owns the intellectual property of the invention and thus for a determined period of time the exclusive right to any financial profit derived from it. The aim of a patent is to allow those who have invested a lot of ingenuity, effort and money into an invention to receive a suitable return on their investment.

Each drug protected by a patent is identified by at least three names. An example is panadol (in the UK) or tylenol (in the US), a well-known drug used to control fever and alleviate mild to moderate pain. Panadol and tylenol are commercial names, and paracetamol is the generic name of the compound, and *N*-acetyl-p-aminophenol or *N*-(4-hydroxyphenyl)acetamide — or shortened acetaminophen — are its chemical names. Drug names tend to be long and often difficult to memorise by the non-chemist.

Are Generic Drugs Trustworthy?

Drug patents last 20 years, after which they expire. From this moment on, any company can produce and commercialise a drug, the intellectual property of which is no longer covered by a patent. Such drugs tend to bear a generic name, for example, paracetamol, but not a brand name. The brand name, for example, panadol, stays the property of the company which initially patented and commercialised the drug in a particular country.

For this reason, drugs with expired patents are called *generics*. This is not really a particularly good name because the adjective "generic" generates the sensation of a low quality product. It implies "low cost", and in our minds everything which is cheap is not worth as much as that which costs more. Especially if we are dealing with a delicate product such as a drug, to which we entrust our well-being.

Yet, generic drugs are on the whole as good as their branded equivalents. By law, they contain the same dose of the same active principle and work by identical mechanisms, yet they cost at least 20% less than branded products. Their safety and efficacy is borne out by about 10 years of wide distribution and ample use of the precursor branded drug. The US law which liberalised the production of off-patent drugs was introduced in the 1980s.

In 2012, 84% of all prescriptions in the US were for generic drugs, whilst the savings accrued by using generics rather than branded drugs in 2011 was 193 billion dollars, an astronomical amount of money!

In most European countries, pharmacists have to dispense the generic version of a drug with a lower price than that of the branded equivalent, and this procedure saves the national health systems undoubtedly a considerable amount of money.

In spite of the excellent credentials of the generics, their reputation is still somewhat shaky. In 2007, researchers at the Brigham and Women's Hospital in Boston, conducted a poll among more than 1,000 patients in order to find out about their opinions on generics. The majority of the participants agreed with the notion that generics are advantageous as compared to branded drugs, less than 10% believed that they cause major side effects. Even though more than half of them believed that Americans should consume more generics than branded drugs, when it came to make the decision for themselves, the percentage decreased to 38%.

In order to combat the distrust in these drugs which is probably linked to their unfortunate name, generics are now often appropriately referred to as *pharmaceutical equivalents*. The causes of this scepticism are of course rooted in issues beyond the name. Ironically, physicians are prominent among those who distrust these drugs. This was borne out by an American opinion poll in which 50% of the clinicians interviewed declared that they had a negative opinion of the quality of equivalent drugs. More than

25% said they had not used these drugs for themselves or their families.

Brand names are easier to remember, especially if they have been around for years. One must of course bear in mind that physicians have to remember several hundred drug names. But one should not underestimate the power of promotion by pharmaceutical companies which may place thoughts into physicians' minds so as to avoid adverse effects on the value of their market shares.

There are also some problems concerning control. For example, 40% of over-the-counter equivalent drugs (those which do not require a prescription) on the American market are produced in India, and the FDA has had some difficulty in supervising pharmaceutical companies on the other side of the ocean. By the time of writing this booklet, the FDA has made amends by introducing spot inspections. They found companies of highest integrity indistinguishable from that of their American counterparts coexisting with firms in which standards of hygiene and quality were quite below Western standards.

In Europe, the national drug regulatory bodies, such as MHRA in the UK, are entrusted with the control and regulation of pharmaceutical equivalents. For cost saving purposes drug production is now more and more relocated to far-away countries, in which regulations are often less strict than in Western countries. Therefore, it seems desirable for the US and Europe to find ways of unifying their regulatory forces so that production and producers of pharmaceutical equivalents can be more effectively controlled and validated in a coordinated fashion, irrespective of where in the world they may be located.

As obligatory controls slowly become routine, the quality of pharmaceutical equivalent production will inevitably improve. This quality, linked to the savings these drugs help accrue, are fundamental to maintain the viability of our precious national health services.

What are Biosimilar Drugs?

Patents of biologicals expire too. Companies other than those which initially patented the innovative product may also produce this type of drug, which is called a *biosimilar* or *follow-on biological*. The problem is the complexity of these drugs.

Two very different drugs, which we have already discussed may help to illustrate this point, acetylsalicylic acid, aspirin, a small molecule consisting of 21 atoms, and recombinant human insulin, a biological made up of 788 atoms organised in 52 molecules. Even more complicated drugs harbour 20,000 atoms. To faithfully duplicate this type of drug means knowing how to reproduce its entire development and production programme, which as we have seen in Chapter 2, includes biosynthetic steps inside living organisms. And all of this has to be done mostly without there being an instruction manual for the complete production process and without the possibility to control it down to the level of each atom.

For this reason, the approval of a biosimilar drug requires a clinical study to be performed in which the new drug is compared with the corresponding branded one in order to confirm safety and efficacy. This obstacle is necessary to guarantee the safety of biosimilar drugs, and it renders their production less attractive for industrial competitors. Therefore, to date there have been only a few equivalent biosimilar drugs.

Shapes and Colours

The dossier required for new drug approval also deals with issues of the outer appearance of the product. The majority of drugs are commercialised in pill form, and you will have undoubtedly encountered a multitude of shapes and colours of pills.

There are companies which deal exclusively with this aspect of drug development. They offer to the pharmaceutical industry the design of an appropriate dosage form of their new drug, for example, in the case of pills or tablets they come up with many geometric shapes and sometimes polymer coatings containing colours which could make an artist green with envy.

Not all colours appeal to patients, and preferences vary with country. In Japan, for example, few patients would take pills coloured dark grey, green or pink. In other countries red, black and colours reminding you of sweets are not appreciated. As the texture needs to be acceptable, drugs are often coated.

But beyond the aesthetic side, coating can also have a functional role in that it can favourably affect the absorption of the active agent or reduce its impact on the wall of the stomach and intestine.

In practice, production of a pill entails mixing the pharmacologically active principle with so-called *excipients* to generate a fine or granular powder. Excipients are substances devoid of any pharmacological activity which possess carefully chosen physical properties ensuring efficient tabletting. The mixture is entered into a tablet press which compresses the powder into the final tablet product.

Tablets are not sold loose but in packages. Also, tastes differ from country to country. Whilst Americans prefer bottles with thirty tablets, the rest of the world is happy with blister packs made of plastic or aluminium.

Whatever the packaging, drugs need to be stable under various conditions of humidity and temperature, and the results of extensive stability tests have to be included in the documentation necessary for obtaining regulatory approval.

Finally in the Pharmacy!

We have arrived at the end of the long journey which has taken our drug to the shelf of a pharmacy, ready to be dispensed to patients who have a doctor's prescription for it.

Each year, more than one and a half million drugs undergo preclinical investigation in the laboratory. About 300 of these make it to the human trial stage, and less than 25 turn out to be promising in phase three studies. In other words, only one new drug molecule in 60,000 makes it! And it takes about eight years from the beginning of the clinical evaluation to regulatory approval.

This long route seems an eternal ordeal especially to patients waiting desperately for a remedy to benefit their condition.

Nevertheless such a prolonged period of clinical investigation serves to spot the many drugs which are inefficacious or indeed dangerous and to eliminate them from the development process. Detailed experimentation is required, because without it we just would not know whether a drug is inactive or dangerous.

Regulatory authorisation is granted for five years, at the end of which authorisation renewal has to be requested. But checks and controls of the effects of the new drug continue beyond authorisation. As we shall see in the next chapter, each new drug released for clinical use is subjected to the so-called *pharmacovigilance* process, an observation phase designed to discover rare side effects which were not picked up in the patients in the original trial of the drug.

Chapter 7
Vigilance, Revisions, New Indications

In 1988, Daniel Simmons, a professor of chemistry at Brigham Young University in Utah, discovered an enzyme which he named cyclooxygenase-2 (COX-2), because it is related to the previously known COX-1, an enzyme which is inhibited by aspirin.

You might recall that aspirin or acetylsalicylic acid blocks the synthesis of prostaglandins by inhibiting cyclooxygenases. Prostaglandins are a group of fatty acids some of which can exert either protective or harmful effects on health.

Simplifying somewhat, one could say that "good" prostaglandins are generated by COX-1, and they protect the stomach lining from the eroding effect of digestive acids. By contrast, COX-2 catalyses the synthesis of "bad" prostaglandins, responsible for painful inflammatory responses in the joints.

Acetylsalicylic acid and other non-steroidal antiinflammatory drugs like ibuprofen interfere with both cyclooxygenases thus reducing inflammation, but they come with some risk of provoking formation of a stomach ulcer. In order to substantially reduce this risk, it is important to ingest such non-steroidal antiinflammatory drugs on a full and not an empty stomach.

The discovery of COX-2 inspired several pharmaceutical companies to embark on the development of drugs selectively hitting this target. The reasoning was that, if COX-2 was selectively inhibited thus causing antiinflammation, COX-1 would remain active and thus the risk of ulcer formation could be eliminated.

Pharmacovigilance Can Dig Up Dishonesty

Rofecoxib has been a product of such a COX-2 inhibitor development programme. In 1998, it entered the US market under the commercial name vioxx.

An estimated 80 million people across the world used rofecoxib, and the aggressive public advertising campaign in the US directed towards the consumer might have contributed to this enormous number. According to the publicity, the drug seemed to be an ideal pain killer without posing any risk to the stomach. In contrast to Europe, in the US advertising of drugs directly to the consumer is allowed, who then asks his or her general practitioner (GP) to prescribe the drug.

Like all new drugs which enter the commercial market, rofecoxib had to undergo the *pharmacovigilance* procedure. This process entails that each adverse drug reaction spontaneously reported by a patient is collected by GPs, pharmacists, local health officials or hospitals. It is then sent to the national pharmacovigilance network organised and supervised by the appropriate national drug agency. The reports are uploaded onto a database and made available to everybody responsible for, or involved with, the study of new drugs.

Pharmacovigilance is important because many side effects, especially rare ones, are sometimes not discovered in the patients recruited into trials in which the drug has been originally evaluated.

In the case of rofecoxib, warning signals appeared from 2001 onwards in the form of infarcts and strokes associated with its use. These ominous signals led in 2004 to the voluntary withdrawal of rofecoxib everywhere by the company which produced it.

Many insiders are of the opinion that the drug should have been recalled much earlier and that in this case the pharmacovigilance process was not sufficiently effective. As it happens, some compromising results emerged subsequently, concerning both drug efficacy and safety, which the company had withheld from the Food and Drug Administration (FDA) when authorisation for Vioxx had been sought.

Pharmacovigilance Can Improve Therapy

Dishonesty is fortunately rare, but much less easily forgotten than positive events which occur more frequently. The pharmacovigilance process identifies rare and previously undetected side effects, and their discovery allows improvement in drug indication. In general, pharmaceutical companies act in good faith and take appropriate steps to improve the treatments they develop. Some drugs can be structurally altered to design toxicity out.

This happened, for example, in the case of a vaccine against rotavirus. This virus causes a form of diarrhoea, which each year causes two million hospital admissions and kills 400,000 children, especially in developing countries. The first vaccine against this virus was approved in 1998, and given the desperate need for an effective treatment, many children were immediately vaccinated. After only a few months, a rare intestinal complication was observed in one in about 12,000 children. This complication had not been observed in the extensive clinical experiments which the vaccine had undergone. After pharmacovigilance generated these warning signals, the vaccine was rapidly withdrawn, and in 2006 a new version was approved and distributed which was devoid of the side effect. Obviously the development of this new version of the vaccine had to restart right at square one. It was tested in a population of children previously unvaccinated, and its efficacy and safety were confirmed. So the approval of each drug is valid for a specific compound showing activity in a selected patient population against a particular health problem. As soon as there is a change in compound, indication or population, one has to start afresh.

Pharmacovigilance Can Discover Resistance

In 1981, the first diagnoses of a mysterious illness which spread rapidly among the gay communities of San Francisco and New York

Figure 1. Luc Montagnier and Françoise Barré-Sinoussi in France and Robert Gallo in the US have independently isolated the HIV virus (with the kind permission of, LEFT, Institut Pasteur — photo Christophe Souber; RIGHT, University of Maryland).

alerted the whole world. Within only two years of the initial diagnoses, the virus responsible, which undermines the immune system to the point of exposing infected individuals to many nasty diseases, was discovered simultaneously by Luc Montagnier and Françoise Barré-Sinoussi in France and Robert Gallo in the US (Fig. 1). This virus, called human immunodeficiency virus (HIV), is sexually transmitted via direct contact of bodily fluids such as blood, or it can be passed on from infected mothers to their babies during pregnancy, labour or delivery. The disease elicited by HIV was named acquired immunodeficiency syndrome (AIDS).

As soon as the virus was isolated, the race was on to find a remedy as rapidly as possible. Researchers were placed under immense pressure by the gay community, initially the most severely affected, which demanded a rapid medical response to this devastating epidemic.

The original idea to develop a vaccine has unfortunately remained a pipe dream to this day. HIV is a champion of camouflage, as it can undergo rapid mutation even within a single infected individual. In contrast, the design of a vaccine capable of instructing the immune system to cope with many possible variants of microorganisms is still beyond the realms of medical sciences.

Fortunately, the hunt for drugs commenced swiftly and turned out to be successful. The US National Cancer Institute took on the task of developing a system of screening all agents contained in the chemical collections of the pharmaceutical industry with the aim to find a drug against HIV. These efforts produced zidovudine, also called azidothymidine (AZT), a compound which after promising evaluation in the laboratory entered clinical trial in more than 10,000 patients at Duke University. Regulatory approval followed shortly afterwards and was enabled by a procedure set up by the FDA called "Emergency Use Authorisation", aiming to stem the lethal epidemic as quickly as possible.

The success of AZT and a little later of other antiviral drugs against HIV gave hope to millions of seropositive individuals. The success could have been of only short duration if the pharmacovigilance process had not signalled in timely manner the first cases of resistance.

Any drug targeted against a microorganism is destined to fail because of the considerable ability of viral or bacterial subpopulations to change their genetic defence make-up, so that they acquire resistance against the drug in at least some patients who receive this type of therapy. It is, however, much more difficult for a microorganism to acquire resistance against a cocktail of drugs administered simultaneously. Today there are more than 34 million individuals who live with the HIV virus. The number of them who die of AIDS each year is in decline, because of the availability of a combination drug therapy called "Highly active antiretroviral therapy" (HAART). This consists of a single pill containing several agents active against HIV, and HAART has reduced deaths from AIDS by 50–70%. Thus it has transformed this disease from a death sentence to a chronic condition. Without pharmacovigilance, the intelligent response of medical research to the emergence of resistance against single antivirals would have occurred much later than it actually did.

For quite a while, drug combination therapies have been in standard clinical use also to overcome resistance against antibiotic or antitumour drugs. When drugs, which have been clinically investigated on their own, are combined with the aim of conquering resistance, the drug development process starts afresh. This means that the combination is subjected to clinical study in order to explore its safety and efficacy in populations of healthy individuals and patients, in spite of the fact that the individual components of the combination have individually already undergone such testing before. Drugs administered together can interact with each other. For example, one component of the combination may increase or slow down the metabolism and elimination of another, with potential detrimental consequences for efficacy and safety.

Indications Beyond Those in the Patient Information Leaflet

When your GP prescribes a drug, you can assume that this drug has been approved for use by an agency such as the European Medicines Agency (EMA) in Europe or the FDA in the US. We have already described in Chapter 6, that approval occurs only if the pharmaceutical company has provided proof that the drug is safe and efficacious in the patient population in whom it was evaluated for a specified medical usage.

Approved medical usages are the so-called *indications*, that is, the specific problems against which a drug has been tested and shown to be beneficial thus justifying its use. For example, the indication for aspirin could be formulated as follows: "Symptomatic treatment of headache, toothache, neuralgia, menstrual pain and rheumatic and muscular pain; symptomatic therapy of fever and flu and cold syndromes". Prescriptions for any other use are *off-label*, which means for indications other than those in the patient information leaflet or patient package insert. But can we be sure that a physician prescribes a drug only for indications for which approval has been granted?

Irrespective of the indications for which a drug has been approved, a medical doctor is allowed to prescribe the drug for any medical purpose, as long as he or she considers such use safe and effective on the basis of his or her knowledge and experience. In the US, one in five drugs are prescribed for off-label use.

Supporters of such off-label usage of drugs make the point that the approval of a drug for a new indication can cost hundreds of millions of euros or dollars. And why should a pharmaceutical company spend such an amount of money on convincing EMA or FDA that a drug might be useful against a non-approved disease, against which it has already been successfully prescribed by some doctors? Or why to spend that amount of money to obtain approval which benefits only a small population of patients who suffer from a rare disease? In this case, costs would be difficult to recover, especially for drugs the patents of which are about to run out, or have already expired.

Whilst off-label prescribing by a physician is legal, a pharmaceutical company would break the law if it advertised off-label use among the medical community. Some US companies have gone down this road, and they had to pay severe fines.

The last in a series of such cases was concerned with risperidone or risperdal, a drug approved in 1993 for the treatment of some symptoms in patients suffering from schizophrenia. Yet, from 1999 to 2005 the drug company which marketed risperidone recommended it also for the treatment of psychotic symptoms such as agitation and impulsiveness in dementia sufferers hospitalised in nursing homes, children with behavioural problems or in patients with mental handicaps. None of these indications has ever been approved. Moreover, several times the FDA rejected requests by the company to extend the indications for the drug. Especially relevant in this context is the fact that risperidone can augment the risk of grave diseases like stroke, especially in the elderly. In spite of the fact that the company knew of these risks, it played them down and continued to recommend the off-label use of the drug.

At the end of a civil and criminal lawsuit, the company was convicted for this illegal practice to pay more than 2.2 billion dollars, an astronomical sum, yet outnumbered by the even more gigantic sales figures. In 2004 alone, risperidone earned the company 3.1 billion dollars.

This case is only the latest in a series which led to at least three other companies having to pay similar fines for related unlawful drug marketing practices. In the case of risperidone the company "... recklessly put at risk the health of some of the most vulnerable members of our society — including young children, the elderly, and the disabled ...", declared the American Justice Minister Eric Holder Jr., when he announced the settlement of the fine to be paid.

In the light of these exemplary convictions, drug experts on the other side of the Atlantic ask themselves how is it possible, that such notices emanate only from the US. Is the European market of these drugs really above reproach? Has no doctor in Europe ever prescribed risperidone or similar drugs for dementia patients in nursing homes or for agitated children or mentally handicapped and difficult to manage individuals?

Drugs Against Non-existent Diseases

There are also examples for these. The story may go like this: A pharmaceutical company identifies a medical opinion leader, let us say a famous physician, who has started talking about a novel disease. Newspapers, TV programmes and internet sites distribute information on this disease, widely implying that there are a large number of people who do not know that they suffer from a severe but curable illness. Curable because a brand-new drug has just come out which cures the disease.

This type of activity has been disdainfully referred to as "disease mongering". A striking example of such practices which actually happened is the promotion of hormone replacement therapy offered to women to "treat" their menopause.

The menopause is a physiological event which happens to all women aged around fifty. The ovaries finish their activity, leading to cessation of menstruation and ending women's fertile age. Although many women experience various disturbances during this period of transition from one to another phase of life, the menopause has never been recognised by medical science as a disease state.

And yet, around the start of the second millennium, women were encouraged to consider the menopause as a risk factor for further diseases, among them osteoporosis or bone weakening, which commonly occurs among women aged 50 or older. It was claimed that such illnesses could be prevented by a hormone-based therapy substituting the very hormones which are no longer produced by the ovaries.

Hormone replacement therapy was suggested to millions of menopausal women, in spite of the inherent risks of cardiovascular diseases, cancer and dementia accompanying it. Companies which marketed the drugs were informed about these risks, but they played them down. Epidemiological studies involving hundred thousands of women who received this therapy demonstrated that the risks outweigh its benefits, and a very acrimonious controversy ensued. Drug regulatory agencies, such as EMA and FDA, instructed the drug companies involved to change the patient package inserts, and to include warnings of the potential risks of the therapy, with the advice to limit its use to controlling specific symptoms for only short periods of time.

The case of the hormone replacement therapy has shocked many, because of the large number of women who were treated for a non-existent disease involving considerable risks. But it has not been the only example.

Drug Pipelines

Predatory strategies in a productive industrial sector such as the pharmaceutical industry are fortunately rare, but they do occur. Apart from generating immediate profits, they are ultimately counterproductive and therefore do not make sense. One has to ask oneself how and why such strategies come about in the first place.

If you have read this booklet thus far, you will realise that it is really difficult to design and develop useful new drugs, i.e. drugs which are efficacious, have only minimal or no side effects and can be prescribed for important medical indications thus benefitting many individuals. Nowadays, the diseases which remain in need of new treatment modes include many types of cancer, disorders of the nervous system and autoimmune diseases. They harbour many biological mysteries, and nobody has yet any convincing idea as to how to treat them all effectively. Even when the pharmaceutical industry succeeds in finding an extraordinarily useful new drug, the road to regulatory approval is riddled with obstacles. So, investment in research and development costs the pharmaceutical industry enormous amounts of money, and potential returns are very uncertain.

It is therefore not surprising, that many big pharmaceutical companies spend a lot of money on advertising those drugs which have already been approved, whilst their so-called drug *pipelines*, that is their research and development efforts generating new drugs, are fairly empty.

Many such companies are off-springs of chemical manufacturing plants founded in the second part of the 19th and early 20th centuries, and after a large number of fusions, acquisitions and take-overs, they have mutated into the few large pharmaceutical multinational companies we know today.

Innovations emanate seldom from laboratories of these companies. Within the last decade, they have actually closed many of their laboratories and transferred research activities to new, much smaller and slicker, enterprises or to academia. So-called *biotechnology* (or biotech) companies sprang up particularly in Silicon Valley, California, on the east coast in the US and in non-profit research centres in several countries.

When a biotech company or a non-profit-orientated laboratory has a promising new drug in its pipeline which has undergone phase one and two clinical evaluation, then large pharma companies tend to get involved. They buy the patent, conduct phase three trials and take the drug towards regulatory approval and ultimately to the pharmacy. Great portions of the budget of these large companies are therefore dedicated to advertising.

The headquarters of the big pharmaceutical industries are mostly situated in the US or in a few European countries. Ever fewer drugs are produced in Western countries, as drug manufacturing units are increasingly being relocated to countries like India, China and Brazil. In these countries, labour costs are much lower than in Western countries, whilst their technical capabilities are rapidly improving.

There is perhaps no economically sustainable alternative to this vastly diffuse panorama of activities, in which a few big capital-rich rather than productive companies dominate the market, whilst smaller firms deal with research and development, and manufacturing occurs in other parts of the world.

Some medical experts believe that the pharmaceutical industry should have a strong moral conscience, and perhaps a completely different business model approach may be suitable: Pharmaceutical industries managed as non-profit organisations, which reinvest their gains wholly into research and development instead of giving them away. A few tentative initiatives leading into this direction have recently been undertaken, mainly, but not exclusively, located in the US. Are non-profit pharmaceutical companies an idealistic and utopian pipe dream, or are they a realistic prospect?

This is impossible to judge at this stage. We need rather to wait patiently for 15 years or longer to find out. All we can do in the meantime is to trust that novel strategies in this area will emerge.

Chapter 8
Predicting the Drugs of the Future

Can we predict how we will treat diseases within the forthcoming decades? "To make predictions is difficult, particularly when they are for the future" is a saying, the attribution of which is uncertain. It might be an old Danish proverb, or it might have been coined by the physicist Niels Bohr, the film producer Samuel Goldwyn or the baseball player Yogi Berra.

Obviously by their very nature, predictions are concerned with the future, and to make correct ones is a challenge. As it is difficult to predict the weather more than three days ahead, it is impossible to know now what the drugs of tomorrow will look like. Bear in mind the above saying whilst reading this chapter. In it, we describe scenarios as to how both the efficacy of drugs and the precision with which they reach their target may be improved, and how the improved understanding of patients and diseases may help reach these objectives. Finally, we shall try to trace a phantom picture of the pharmacologist of tomorrow. Do not expect certainties from this chapter, questions concerning the future are much more numerous than any answers.

Nanovehicles for Drugs

"We have many excellent medicines, but lack good taxi drivers". This remark is by Mauro Ferrari, a bioengineer specialised in nanotechnology and director of the Methodist Research Institute in Houston, Texas. He designs tiny vehicles capable of carrying drugs only to where they should get to, avoiding them getting anywhere else. Selective distribution of drugs is one of the unresolved problems of contemporary pharmacology, as we have said many times in this booklet.

We already do know how to guide a drug to certain parts of the organism, for example, by attaching it to a compound called an *antibody* which functions as a courier. The idea is that the antibody carries the drug only to those cells which have the corresponding antigen on their surface. This concept has been exploited in experimental tumour therapy. If the antigen which reacts with the drug-carrying antibody was only located on tumour cells, things could work out nicely. Up to now though, the results of experiments with drugs conjugated to antibodies have turned out somewhat disappointing. A reason might be that the antibodies are insufficiently specific for tumour cells. Another difficulty associated with this approach is that the chemical link between drug and antibody must be stable enough to allow transport of the drug to the tumour, but at the same time sufficiently unstable to permit the eventual release of the drug from the conjugate once it has entered the diseased cells. One current method to improve drug distribution in tissues in which one wants them to exert their activity is to encapsulate them in tiny vehicles, the size of which is in the order of magnitude of nanometres (a nanometre is a billionth of a metre). Such vehicles need to be designed in such a way that they carry the drug to the correct locus of desired action. Iron nanoparticles, for example, can be used as vehicles to transport drugs into tumour cells characterised by the ability of absorbing iron more efficiently than healthy cells do.

Nanotechnologies are all the rage just now, and the expectation is that they will revolutionise both pharmacology and medicine, as they have already impacted cosmetics and the design of materials used in sport and technical clothing. Expectations with respect to nanopharmaceuticals are high.

Examples of drug formulations involving nanoparticles already exist, such as small lipid spheres capable of incorporating drugs. These so-called *liposomes* maintain the activity of the incorporated drug but show reduced toxicity, thus offering a therapeutic advantage over the drug when formulated and administered in the traditional way. The development of different types of nanoparticles using biocompatible and non-toxic materials offers considerable new opportunities to the field of pharmacology, and many research laboratories all over the world work in this area. One could foresee the possibility of designing drug-carrying particles capable of penetrating the brain, which as we have said in Chapter 2 is more or less inaccessible to many drugs. And there is the prospect of designing drug carrier materials which reach certain tumours at substantial concentrations and release there the drug they carry at the rate and quantity required for antitumour activity.

The reason for the attractiveness of nanopharmaceuticals is that they can be constructed with a variety of building blocks allowing modularisation of their physicochemical and biological properties according to their ultimate purpose. Furthermore, nanoparticles may act as vehicles for drug molecules which cannot be administered in the traditional way, because of poor stability, as for example, RNA molecules which act by blocking the expression of diseased genes responsible for illnesses.

Are nanopharmaceuticals safe? This is an area which needs still a lot of investigation. It seems that because of their tiny size, nanomaterials are more likely to obey the laws of quantum mechanics than those of thermodynamics, both of which influence the behaviour of matter in our micro- and macroscopic worlds. In order to understand even in only approximate fashion, the difference between these two worlds, we take as an example the material gold. When occurring at normal dimensions, gold resists destruction by most chemicals, even aggressive ones. Stability and durability constitute most valuable assets of gold, and they have been appreciated already in ancient times. These properties have been exploited in medicine, think only of the gold crowns which are still being used to replace damaged parts of teeth.

In contrast, nanosized gold does not possess the properties of gold which we are familiar with. For example, the colour of nano gold is red rather than yellow, and its stability is not always the same as that of larger sized gold particles. Results of several toxicological studies have hitherto been inconsistent. It appears that the characteristics of gold nanoparticles when immersed in a physiological solution which resembles human blood do not remain constant, as is the case with gold of larger particle size.

The take home message here is that, nanopharmaceuticals are extremely promising but they — just as all new drugs — need to be studied further very carefully, in order to establish whether they are safe as well as efficacious.

Selecting the Right Patients for the Right Therapy

An objective which contemporary pharmacology has hitherto not sufficiently focussed on is, how to administer the right drug at the correct dose and the right dose schedule to the right person for the right disease. This approach is referred to as "precision medicine", and it is currently a hot topic among health professionals. But all these "rights" and "corrects" do not signal that the objective is easily achieved, because it requires the intimate and detailed knowledge of activities and side effects for each drug in all individuals and all diseases. However, tentative attempts have been made to improve the match between drug, disease and patient.

For example, some ovarian tumours express high levels of a receptor for folic acid (also called vitamin B9), which picks up folic acid and carries it into the cell. Cognisant of this fact, scientists have thought of linking an antitumour drug to folic acid in order to help it selectively penetrate ovarian tumour tissue. The problem is, however, that not all ovarian tumours possess high levels of this folic acid-binding receptor, so that only some patients would respond to this therapy. To help resolve this conundrum, a diagnostic test has been developed which allows prospective evaluation of those patients whose tumours have sufficient receptor expression levels. This test allows the selection of those who have a good probability of benefitting from the drug. Another example which has already been discussed in Chapter 4 is herceptin, which is administered only to those patients whose breast tumours produce high levels of the receptor Her2.

It is conceivable that this type of therapeutic approach will become more prominent in the future, in which drug administration is preceded by a diagnostic test allowing *a priori* the selection of the patients whose disease is sensitive towards the action of the drug. Therapies can thus be paired with precise molecular diagnoses, which not only characterise the patient's disease but also help select the most appropriate therapy and predict the patient's response. Another related approach is to administer drugs after patients have undergone genetic tests to find out whether they harbour DNA variants which render them sensitive to the action of the drugs. Tests of this type are already being explored for a number of tumour types.

New Indications for Old Drugs

You surely have come across examples of old buildings of now disused factories, which have been converted into marvellous new exhibition or living spaces. Their outer appearance is unchanged from what they looked like originally, what has changed is their use. This analogy holds also for some old drugs, which after years of relative obscurity in "retirement" undergo what is termed *repurposing* (or *repositioning*), because they are unexpectedly found to offer novel benefits to patient groups different from those in whom the drugs were used originally.

In other words, drug repurposing is the application of known drugs to new indications. In Chapter 6, we have hinted at the second life of thalidomide, a drug which after having caused a tragedy because of its toxicity, was salvaged and given a new lease of life. Decades after thalidomide had been withdrawn from the market, it was found in the late 1990s to be capable of killing multiple myeloma cells. Given that thalidomide had already been pharmacologically evaluated and received regulatory approval, the drug could be tried out immediately in myeloma patients. It was found to be a reasonably active therapy against a tumour which up to then had been very difficult to treat. Since myeloma occurs predominantly in middle aged or older people, in this therapeutic scenario thalidomide does not carry the risk of toxicity in the offsprings of pregnant women, the reason for the erstwhile thalidomide withdrawal.

Studies conducted in the past five-to-ten years allow preliminary clues as to how thalidomide may act against myeloma. The drug causes the loss of two transcription factors, i.e. protein molecules that bind to specific DNA sequences thereby controlling the rate of transcription of genes, important in this case for tumour maintenance. Their loss provokes tumour growth arrest and modification of the immune system.

Sometimes the knowledge of the mechanism of action of a drug leads to the realisation that the biomolecules or molecular circuits targeted by the drug are important not only in one but in more than one — apparently unrelated — disease states. A case in point is metformin. Since 1958, this drug has helped diabetics to lower their blood glucose levels. Several years ago it was discovered that metformin might also prevent the development of certain tumours.

The antidiabetic action of compounds such as metformin has been known since ancient times, given the fact that the plant *Galega officinalis*, also known as goat's-rue or French lilac, contains compounds chemically related to metformin (Fig. 1). Extracts of Galega have been used since the middle ages to control blood glucose levels, and studies dating back to more than 100 years ago of the constituents of this plant contributed considerably to the development of this type of antidiabetic drug. Metformin reduces the production of glucose in the liver and augments its consumption by peripheral tissues. It is today the world's most frequently prescribed antidiabetic drug, not least due to the fact that, as a very old drug, it is cheap.

Figure 1. *Galega officinalis* produces a substance similar to metformin, which helps control blood sugar (Ulf Eliasson, via Wikipedia, under the Creative Commons Attribution 3.0 Unported license).

Apart from controlling blood glucose levels, metformin has been shown to interfere with tumour cell proliferation in several types of cancer in laboratory experiments and in preliminary trials in humans. These findings have highlighted the fact that some pathways germane to human metabolism, generating energy from nutrients, are equally involved in diabetes and cancer.

How these pathways may augment cancer development, and how metformin may disturb them to prevent certain cancers, is unclear and the topic of intense current pharmacological research activities. If the anticancer activity of metformin in humans will be confirmed in future research, it might be possible to use it not only as an antidiabetic drug but also as an agent to prevent certain cancers, for example, in overweight individuals, who are at an increased risk of both diabetes and certain cancers.

Other examples of repurposed drugs which received regulatory approval long ago are some employed in drug combinations. As we have discussed in Chapter 7, in order to combat certain diseases, one single therapeutic "weapon" is not enough. A drug which on its own is devoid of activity or only weakly active may well exert impressive activity when combined with another drug. Combining two or more drugs is an appropriate strategy when the activity of the combination is superior to the sum of the pharmacological effects of each single component. Such drug combinations can be particularly apposite to overcome drug resistance, as borne out by the drug combinations currently used against acquired immunodeficiency syndrome (AIDS) or bacteria with acquired resistance against single antibiotics.

In the future, we may well see many more examples of "old" drugs, i.e. those which have been used for a long time, undergoing repurposing for novel usages and indications. As we gradually improve our knowledge of the metabolic pathways and molecular circuits targeted by these old drugs, we may well stumble across multiple overlaps between diverse diseases. In this way, opportunities for the innovative use of some old drug molecules may arise. There may well be many age-old and almost forgotten drugs languishing on the shelves of the pharmaceutical industry, the hitherto untapped pharmacological potential of which is waiting to be discovered and to surprise us.

Biobanks — Sources of Priceless Treasures

Fossilised residues of a human who lived several million years ago can tell us a lot about the origins of human life. In the same way, a biological sample originally derived from a patient suffering from a particular disease and stored in a hospital biobank, i.e. a repository of human samples, can provide important information about the nature of the disease and on how a patient who is afflicted by the disease might be optimally treated.

Here is an example. Imagine a laboratory discovery suggesting that drug X, which has been used for many years to treat patients with a particular disease, is more efficacious in a subset of the disease which is characterised by a certain molecular characteristic, than in the disease when it lacks that property. The best way to verify whether this finding applies also to patients is to conduct a clinical experiment, as outlined in Chapter 5. Specimens available from patients who in the past have been treated with drug X are thus investigated to furnish initial insights which can help decide whether it is worth embarking on a long and expensive clinical study. In practice, specimens are taken from the biobank and analysed as to whether they possess the molecular characteristic under study or not. As each specimen should come with documentation describing the clinical history of the patient from whom it was taken, it is possible to establish whether our hypothesis formulated in the laboratory is confirmed in human samples. Thus, it is possible to find out whether patients with the disease characteristic under study had a better clinical outcome after treatment with drug X than those who also received drug X, but whose disease lacked this molecular characteristic. Clinical studies on archived human specimens are called *retrospective,* because they inspect events in the past, whilst studies into which patients are recruited to explore the effects of a new drug are called *prospective,* because they investigate events which have still to occur.

For retrospective studies molecular biological investigation methods such as gene or protein analyses are used today, which did not exist when many of the specimens were obtained and archived. The biological insights into diseases and effects of treatments increase constantly in parallel with the introduction of more and more sophisticated techniques. The possibility of accessing specimens archived long ago is therefore an extraordinary resource which helps to test hypotheses rapidly and economically and is aimed at improving the treatment of patients who today present with the same problems as they did many years ago.

Many of the patients who consented to their specimens being used for research purposes are no longer around, but their tiny biological donations, which may have seemed so trivial at the time, are really a tremendous gift to patients today and tomorrow. The value of these samples may even be greater if and when the dissemination of the information which they provide will be optimally organised, so that as many medical researchers anywhere in the world as possible can make good use of it.

The original motivation for archiving patient specimens in the hospital was the legal obligation aimed at allowing resolution of potential disputes arising from treatments. Today such specimen collections have the potential of becoming biobanks of great scientific value. These repositories can serve as intelligent archives if explored in the right way to tell us a lot of things about a disease and its potential treatment. However, there are several conditions which have to be met to render such biobanks more useful than just freezers! A biobank is valuable if apart from conserving specimens under optimal conditions, it has been set up in a carefully thought-through fashion; if each sample is preserved in the appropriate environment which avoids degradation and allows easy access and retrieval for research; if the clinical history of each patient who served as donor has been stored meticulously in the accompanying database; and if each patient has given the appropriate consent for the use of his or her specimens for research purposes.

The tremendous effort which is involved in setting a biobank up is completely wasted without accompanying robust informatics programmes capable of organising, managing, reading and interpreting data. The software involved must be able to provide comprehensive information on the physical location of the specimens, the clinical history associated with each sample pertaining to both disease and treatments received, the scientists who have been allowed to access specimens, the type of study which they performed, the time frame in which this was conducted, and the results which emerged from it, and so on.

All this information can be a gold mine for future discoveries. If it has been obtained in reliable fashion, is described in precise terms and is well-organised, it can help characterise patients in terms of their diseases and the effects of their treatments. Informatics programmes which can handle this information are not written or used in isolation, rather they require the involvement of individuals familiar with the many regulations and the multitude of molecules encountered in drug research. We are talking here about the so-called bioinformatics specialists.

Digesting Big Data Sets

The aim of the bioinformatics discipline called "systems biology" is not difficult to understand, yet it is difficult to achieve. It is to sift through great amounts of data searching for unexpected features of biomolecules such as genes or proteins and of their interactions in cells, tissues or whole organisms. Systems biology is an engineering approach which entails simultaneous measurement of the dynamic network of multiple biological interactions, without speculatively focussing *a priori* on a specific relevant biological variable or principal circuit. Systems biology supports large scale projects like the Cancer Genome Atlas or the "omics" projects which we have touched upon in Chapter 2.

However, bioinformatics is unlikely to provide a magical solution to each biological mystery. One of its most eminent critics is the Nobel Prize winner Sydney Brenner who maintains that "we are drowning in a sea of data whilst being starved of knowledge". Brenner is particularly critical of the practice of collecting biological data without a solid hypothesis having been formulated *a priori*. Timothy Hunt, another Nobel Prize winner, is equally sceptical. He sees it like this: "A lot of people think that if one measures absolutely everything, the truth will jump out like an eyesore... Yet experience teaches us that it is more a recipe for confusion than anything else".

Are the two Nobel laureates right, or are they perhaps a little too sceptical? Certainly, results of bioinformatics studies need to be evaluated with great caution and considered strictly for what they are. Bioinformatics can be a powerful tool in terms of enabling the analysis of many factors together to come up with correlations, for example, between the frequency of a certain gene on the one side and a certain type of disease on the other. But the two factors might also be associated by chance, without one influencing the other or *vice versa*. You may recall that in Chapter 5, we have already mentioned the difference between a link between cause and effect and a simple correlation.

Another problem is that the number of correlations discovered by bioinformatics analyses can be enormous. How can one then distinguish the ones which are important from those which are not? Only traditional laboratory experiments in which the role of each and every molecule in the cell or whole organism is examined can really verify inferences drawn from data thrown up by bioinformatics analyses.

If the great databases contain errors, they tend to be multiplied, just like entries in Wikipedia: Such entries are frequently cited in essays and books, and so erroneous information can be disseminated. This process is comparable to the propagation in the scientific literature of flawed but popular and credible scientific results when restated by scientists who cite them.

Ultimately, the software which searches for information within great quantities of data functions better if the phenomenon searched for is frequent, but it is prone to mistakes if the phenomenon is rare. For example, Google Translator is capable of translating in reasonably correct manner from or into English, because it has amassed a lot of experience with this language, one of the most frequently spoken in the world. This is much less the case with regards to Google translation from or into a less popular language such as Italian. In analogy, a bioinformatics study which analyses data concerning a well-known disease from which many patients suffer produces more reliable results than a study which scrutinises data derived from a rare disease afflicting only a few patients.

Therefore, the take home message is: bioinformatics methods can be a powerful research tool if used appropriately, considering carefully their limits. Be warned not to expect magical insights from bioinformatics!

Simulations and Imaging

Information derived by bioinformatics approaches can be exploited not only to conduct analyses but also to simulate and visualise biological processes. Nowadays, sophisticated programmes allow at least partial reproduction of events occurring in molecular and cellular circuits when the simulation is based on large amounts of real data accrued in laboratory and clinical experiments. It is thus possible to formulate hypotheses concerning the appearance and shape of biomolecules implicated in disease causation and their interaction with drugs.

Such simulations, which scientists call *in silico* because they are performed on the computer and not directly in laboratory experiments, tend to be rather crude. They focus on a small portion of a biological molecule and reproduce in most approximate fashion the real cellular scenario. Many scientists believe that it is just a matter of time that such simulations will become even more reliable representations of biological events.

Some of the more mature readers might recall the times when computers attempted to reproduce common objects. Such objects were outlined by a few rough strokes on the screen, and the resulting images bore really little resemblance with reality. In contrast, today it is often difficult to distinguish between an animated digital cartoon and a screenshot with actors in flesh and blood. The gigantic advances in this technology are obvious to everybody, similarly perhaps, the quality of biological computer simulations might undergo a comparable quantum leap within the next few years.

Drugs Which Repair Rather Than Treat

Will it ever be possible to eliminate the Human immunodeficiency virus (HIV) virus from all cells of a seropositive AIDS sufferer? Or to eliminate all detrimental genetic alterations from the cells of a cancer patient? Or in other words, will we ever be able to achieve the goal of designing drugs which repair, not only treat, diseases?

Let us start with the possibility of curing AIDS. This seems a suitable example because the goal of repairing, although remote, seems nowadays somewhat more attainable than it was only a few years ago.

For years efforts to eradicate each bit of the HIV virus from the body of a seropositive person seemed beyond our medical capabilities. The virus, which infects a human, replicates within millions of cells of the human immune system thus depositing its most precious possession, its genetic material, inside our most precious possession, our DNA safeguarded in the cell nucleus.

As the infection proceeds, the virus slowly continues to reproduce itself and infects more and more new cells, penetrating in ever greater numbers cell nuclei and depositing there ever more copies of its genetic material. Therefore, to achieve a complete cure would require the complete elimination of these deposits from the nuclei of each infected cell. Mission impossible?

Perhaps the mission is indeed impossible, at least at the moment. But let us try a different tack. Is it, instead of attempting to eliminate the virus, perhaps possible to eliminate the entry door which the virus uses to get into the cell?

In order to penetrate into cells, more or less all animal viruses use receptors located on the cell membrane as the entry door. HIV uses at least two such protein receptors in the majority of individuals, one is called CD4 and another CCR5.

Whilst one can hardly survive without CD4, we can do without CCR5. This has been demonstrated by the very few fortunate patients who possess a mutation in both CCR5 genes. They are healthy and immune to AIDS. Individuals who are homozygous with respect to this mutation tend not to be infected by HIV even on repeated exposure to the virus. This is because in their body HIV cannot find the entry door to penetrate into cells and infect them.

The story of one of the very rare patients who was definitely cured of HIV confirms the importance of the receptor CCR5. Timothy Brown is also known as the "patient from Berlin", as his cure was achieved in the German capital. Brown had AIDS and suffered also from leukaemia. One of the therapies against his blood tumour was a bone marrow transplant. The bone marrow is a soft tissue located within the bones, where blood cells are formed.

The idea of the treating physicians was probably: "Let us try and kill two birds with one stone. As we have to perform the transplant, why not try and substitute Brown's bone marrow cells with healthy cells which are also resistant against HIV". That might have been their thinking.

Thus a bone marrow donor was chosen who was resistant against HIV because of a homozygous mutation in his CCR5 gene. Brown received the transplant in 2008, and from then on he was healed of either disease. He also does not need to take antiretroviral drugs against HIV.

What might have happened to his organism? The transplanted cells which lacked the receptor CCR5 and thus were resistant against HIV infection were strong and healthy and probably got the upper hand. They might have reproduced themselves with more vigour than Brown's own battered infected cells still present in several marrow organs such as lymph nodes and spleen. It is conceivable that the infected cells did not survive in the marrow, which had been irradiated prior to the transplant.

Brown's recovery has raised the hope that other patients can be cured of AIDS if it is possible to eliminate the receptor CCR5 from their cells. However, marrow transplantation is a complicated and expensive procedure, and immuno-compatible donors who also have the homozygous mutation in the CCR5 gene are extremely rare. So we may have to pursue another avenue.

Imagine a tiny genetic surgeon who can enter into a cell, penetrate the nucleus to cut out bits of the CCR5 gene rendering it useless. Such a micro-surgeon who is able to confer resistance against HIV on the cells of the immune system does indeed exist. It is an artificial protein with the slightly off-putting name *zinc finger nuclease*.

Do not be turned off by this unfortunate name and read on. The artificial protein has been designed such that it contains an enzyme (the nuclease) capable of cleaving DNA and a protein (the zinc finger) which can ligate DNA sequences, all of this happening in a precise position of the genome. In our case, the zinc finger nuclease has been designed to cut out specifically the gene CCR5.

The approach has worked well in the laboratory and to some extent in patients. White blood cells were removed from patients' blood and treated with a zinc finger nuclease directed at the CCR5 gene. The enzyme reached the nucleus by way of infection with another virus, an inactive adenovirus. After treatment, the white blood cells had acquired resistance against infection with HIV, and after reinfusion into the patients' blood, the virus load present was reduced substantially, although not eradicated.

In the light of the fact that the substitution of diseased cells was not complete and many unmodified cells could still be infected by HIV, this encouraging result is still far from engendering a new cure for patients. The use of an inactivated virus as vehicle to penetrate cells and to access the nucleus carries risks which need to be considered.

Nevertheless, the result constitutes a most important proof of principle. It has been achieved in the laboratory of Carl June at the University of Pennsylvania — make a note of his name, which may well some time crop up among those of the prize winners in Stockholm! This proof of principle not only shows tremendous scientific imagination but also gives hope to patients.

Dreaming of the Molecular Instruments of Tomorrow

In the not too distant future we shall perhaps be able to package devices such as the synthetic protein used by Carl June into a drug. Today such devices can only be used in an experimental setting, by extracting cells and manipulating them outside the body before they are reinfused. Maybe a drug of the future will be able to carry devices, like the genetic micro-surgeon described above, into the nucleus of all cells which require repair, for example, making use of a specifically designed nanovehicle, without the need for a modified virus for transport which always comes with the potential of unwanted effects. In the case of HIV, such a drug of the future may be able to destroy both copies of the CCR5 gene in each infected cell. Perhaps medicines will be developed which can cure in analogous fashion other diseases such as cancer, the maintenance of which depends on cellular genetic lesions.

Or scientists may achieve the ability to manipulate not only enzymes or antibodies but also much smaller molecules which our body already uses to control cellular events dynamically. The exploration of such molecules and processes has already started, and scientists have given them exotic names such as "microRNA" and "epigenetics" among others. Ignore the names because the concepts are fairly simple. Our DNA is the same in all cells, but cells differ from each other and perform different tasks. This diversity is regulated by molecules such as microRNAs, whilst epigenetic phenomena determine in each cell which parts of the DNA are read and used.

Today the idea of drugs containing microRNAs is still a pipe dream, but dreaming can be productive. It can help to imagine what one may have to do to overcome the obstacles on the way from the drugs of today to the innovative ones to emerge tomorrow.

Who Will Invent the Drugs of Tomorrow?

We would like to think that some young readers of this book may by now have become so inspired and enthusiastic that they may consider embarking on a career dedicated to the invention of new medicines.

A question which adolescents in this situation may pose is this: "Which is the best faculty at which I should enrol to learn how to design drugs?" Let us say straight away that there are few wrong answers to this question. Sure, reading ancient literature or archaeology may not be the most direct avenue towards reaching this aim, but the choice of relevant subjects to be studied is enormous, and most good scientific faculties of universities and colleges can groom pioneers in this field.

Designing new drugs requires many diverse proficiencies. We need surgeons, medics and pathologists who know about human diseases and their complicated manifestations. We need biologists who understand how to unravel the multitude of circuits and biomolecules which make each — even minor — activity of our body function correctly or indeed incorrectly. We need chemists and physicists who know the laws of matter, even in its most minute manifestations, and are able to manipulate them. We need pharmacologists who know about the thousands of drugs of yesterday and today, and their properties and side effects. We need veterinaries to develop models of human diseases in animals, our indispensable allies in the research for new cures. We need engineers who can invent small or large instruments and machines to encapsulate and transport drugs. We need mathematicians and statisticians to help calculate the appropriate number of study subjects and interpret results from laboratory and clinical studies. We need informatics specialists to organise the information gained and the processes used in such studies and to simulate encounters *in silico* between drug and human organism before the insights gained by such simulations can be transferred into clinical reality. We need research managers able to digest the ever increasing organisational complexity of the drug development process and of the appropriate experimentation. We need psychologists who can work with patients and explain to them what might happen in a clinical study to which they wish to make their precious contribution.

This list is undoubtedly deficient, because of our lack of imagination, given that the expertise required to design the drugs of tomorrow might well be even more diverse than what is included in the list. The one important piece of advice which we think we can give confidently to those who might be attracted to take the adventurous road of drug research is to choose the discipline which seems to you the most exciting one to embark on. If there is sufficient interest, passion, tenacity and courage, the best path can undoubtedly be found! What is important is to choose good and inspirational teachers, and to be prepared to make sacrifices and not worry too much about having to go to perhaps anywhere in the world.

Chapter 9
Not Every Pill is a Proper Drug

We are almost at the end of our journey across the world of the imperfect science of drug development, taking you from the original idea all the way to the packaged vial. We have seen that the original idea for a new drug arises from an unmet medical need. From this idea emerges ultimately a compound, the potential new drug. It may be an agent which exists already in nature and can be manipulated in a chemical laboratory to improve its stability or tolerability, or to enable an easy chemical production method. Alternatively the compound is newly synthesised in the chemical laboratory. Or it may be the product of the synthetic capacity of a cell, perhaps a genetically modified one, which constitutes a most proficient chemical laboratory, capable of generating innovative drug molecules such as the monoclonal antibodies and other biologicals in use today.

Whatever their origin, each of these promising but initially unproven new compounds needs to be subjected to studies first in the laboratory and then in the clinic, before they reach ultimately the stage of authorisation and regulation germane to new drug approval. As the general public expects convincing evidence of not only efficacy but also safety, most new drug molecules fall by the wayside during these investigations. Only very few are successful and reach the end of the route.

The strict rules in force to guarantee the safety of new drugs often slow new drug development down, even if the drug under study might have the potential to cure a hitherto untreatable disease, and some doctors and patients might therefore be prepared to accept a certain risk of toxicity.

Yet, the high number of new drugs which fail at some stage of their development process has always been only in part caused by the rigorous rules imposed by new drug-approving bodies. The most important reason for failure has been and still is the fact that the extent of our knowledge of the functions of the human body, its diseases and their cures, resembles more the small crest of a wave rather than the whole ocean. Most disease details and mechanisms have yet to be discovered.

Our profound ignorance is also the reason for having to interpret results of each clinical study using statistical methods. If we knew right from the outset that a drug at certain doses and dose schedules was safe and efficacious in a particular patient cohort, the results to emerge from such studies would be much neater and tidier than they tend to be in reality. Then it would not be necessary to resort to probability considerations in order to exclude the possibilities that the drug trial is conducted in an inadequate number of individuals, or that the study results are either wrong because of error or correct because of chance.

Our limited and imperfect knowledge is compounded by the awareness that there is not one person in the world who is the same as another and that any drug can produce different effects in different individuals. This realisation makes drug researchers go about their business with great caution, even in the case of agents which have been successful and granted approval. Utmost vigilance is appropriate even with respect to drugs which have been sold in pharmacies for years and used by much larger numbers of people than those who were involved in the original clinical evaluation.

Sometimes, pharmacovigilance discovers that a drug has not been experimentally evaluated in the way it should have been, which entails serious implications often involving judicial proceedings. In other cases, pharmacovigilance engenders positive surprises, in that a drug designed to treat one health problem is shown to unexpectedly have the ability to cure another.

Cheating and scandals involving the development of new drugs stay in one's memory much more vividly than the circumstances of the discoveries of the many drugs which have been safe, active and extraordinarily useful, and gone on to improve medicine. Some arbitrarily chosen but noteworthy examples are aspirin, penicillin and gleevec. If you are still not convinced that drugs have made our lives more agreeable than those of our ancestors were, try to imagine what life would be like without drugs.

Imagining a World without Drugs

Let us take a large leap into the past and imagine we are in a primitive operating theatre at the beginning of the 19th century, where patients arrive on the surgeon's table inebriated with alcohol and other agents which knocks them profoundly out. Not surprisingly operations were rare and restricted to interventions which offered at least a little hope of success, given the atrocious pain and the large amounts of lost blood involved.

Extreme speed was of the essence for these ancient operations, which had remained basically unchanged from times of antiquity. The amputation of a limb could be completed in less than a minute, this being the only way to avoid the poor patient's fate of immeasurable pain.

To escape the marginal treatment niche which surgeons occupied in those days, three innovations were crucial: Antibiotics and disinfectants to stop infections, drugs and instruments to control bleeding and anaesthetics to inhibit the feeling of pain. We have already talked about the control of infections and blood flow, yet a tribute to anaesthetics seems appropriate here.

Before the advent of anaesthetics, the desperate cries of patients filled ancient operation theatres followed by telling silence. Today, one hears cries of pain only in delivery rooms of hospitals in which the use of anaesthesia for women in labour is still discouraged.

Without anaesthetics, organs like the abdomen, thorax and joints, not to mention heart or brain, would remain out of reach of surgical procedures. At one time, such procedures were only conducted on the most external parts of the body, and the rest was left to specialists in so-called *internal medicine*. Routine operations performed every day in hundred thousands of patients, such as coronary artery bypass, graft surgery or replacement of a damaged hip with an artificial one, would be unimaginable.

Anaesthetics were used for the first time in the middle of the 19th century in the practice of an American dentist, William Morton in Boston, and then transferred rapidly to operating theatres everywhere. Anaesthetics are only one example of so many drugs, not only life-saving ones, which have changed human life. Yet not all pills are proper drugs.

Pills Which Are not Drugs

Vitamins, tonics, food supplements, antioxidants, slimming aids, herbal extracts, Bach flower remedies, *Ayurvedic* medicines and homoeopathic remedies — the shelves of pharmacies and drug stores are crammed with products of this type. They possess certain properties of medicines, their packaging, shape and description of presumed curative properties, but they are certainly not proper medicines (Fig. 1). In some cases such *"parapharmaceuticals"* can even be dangerous.

They arrive in the pharmacy, without having passed the stringent controls which proper drugs have to undergo, as we discussed in Chapters 3 and 7. Parapharmaceuticals do not have to be submitted to laboratory tests or clinical investigation, and there are no pharmacovigilance procedures in place to monitor adverse events after they have been introduced into the market. They may contain detrimental excipients or other substances which differ, or are absent, from those listed in the information leaflet or on the label. Moreover, ingredient composition or dose may differ from one batch of such products to another, because the law does not require that they undergo stringent pharmaceutical quality control procedures.

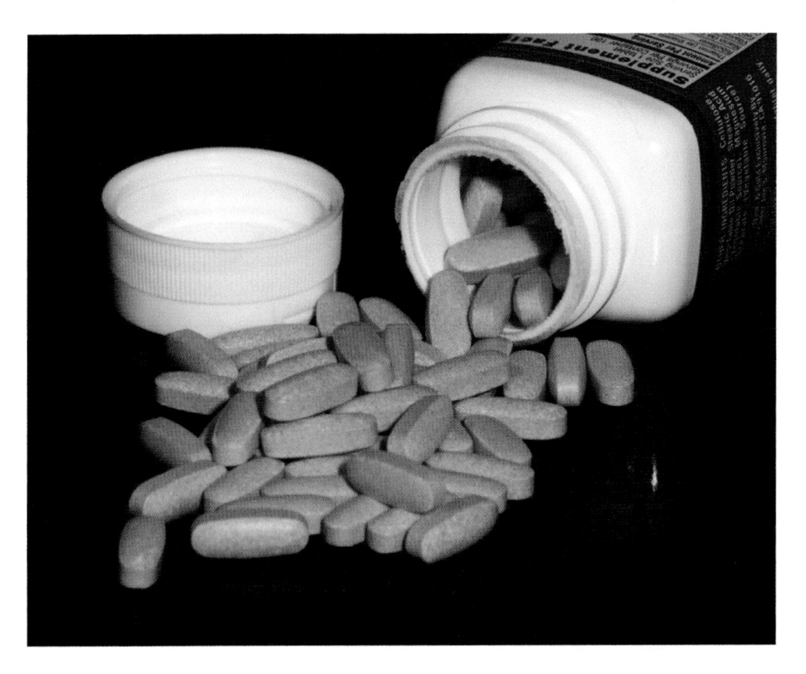

Figure 1. Supplements: what they contain is not necessarily known and they can be quite dangerous (Ragesoss, via Wikipedia, under the GNU Free Documentation License).

Some of these parapharmaceuticals are available not only in pharmacies and drug stores but also in herb stalls, supermarkets, fitness studios and shops for sports articles, where shop assistants do not have the thorough training required to adequately advise clients. Even more dubious are "online pills", health products which are advertised and purchasable on the internet. Here the potential risk of obtaining counterfeit products which may be harmful as they have not undergone any safety controls is considerable (Fig. 2).

Especially in the case of vitamins and nutritional supplements there is really no rational reason to ingest such products, most of which have never been either studied in experiments or subjected to safety controls and thus may be inactive and/or detrimental to human health. A varied and balanced diet, rich in fruits and

Figure 2. The platform where canisters containing pharmaceuticals are loaded into an automatic dispensing machine at a mail order pharmacy (US National Institute for Occupational Safety and Health, via Wikipedia).

vegetables is more than sufficient to supply our organism with everything it needs.

The Odd World of Homoeopathy

In medicine, there is a general rule which says that the greater the dose the greater the pharmacological effect, and this has been more or less demonstrated by a century of controlled clinical studies. Like all good rules, this one is also not without exception. In recent years, interesting *hormetic* phenomena have been discovered in pharmacology, showing convincingly that selected drugs or poisons at very low (but still measurable) doses or concentrations can exert certain biological effects which cannot be observed, or are different from those seen at higher doses or concentrations. In contrast, homoeopathy goes against any scientific reasoning, suggesting that there are active principals which, in order to be more efficacious need to be diluted to the point at which no original active molecule can be left in the vial. As a consequence, homoeopathic products tend to contain excipients and hardly any active agent at all.

The practical absence of active principle has impeded experiments to be performed on homoeopathic products, which tend to be delivered to the pharmacy without any verification as to their exact contents and their presumed efficacy and safety, because they contain more or less exclusively sugar (in the case of globules or tablets) or dilute ethanol (in the case of droplets).

One should not confuse homoeopathic products with the co-called *phytotherapeutics*, which are extracts of plants containing pharmacologically active principles, although their efficacy and safety may well not have been tested or proven.

Supporters of homoeopathy take advantage of the "absence of anything" in homoeopathic products, except sugar or aqueous alcohol, to support the claim that, whilst these remedies may not be quite as active as conventional drugs, they are also certainly not as toxic. This is a rather illusory claim. One just cannot state that a drug is able to exert either benefit or detriment in patients suffering from a given disease without having tested this drug thoroughly. If a sick person who has the need for treatment takes a homoeopathic remedy instead of an allopathic drug which might help improve his or her condition, his or her problems may be severely aggravated and even death may occur, because the treatment is inadequate.

Perhaps, these issues are not thought through carefully by the staggering number of patients in Europe who consume homoeopathic remedies in ever greater numbers.

Be Careful!

Before saying goodbye to the reader, there are a few recommendations we think we have to administer. If you have read up to here, you will realise that drugs can be extraordinarily useful, but that they are objects which have to be handled with utmost care.

Drugs should be taken at the right dose and the appropriate dose schedule only if there is a real need, and when they have been prescribed by a medical doctor. One should absolutely refrain from any self-medication or from taking drugs recommended by a friend or relative, unless the friend or relative is a medical doctor or pharmacist.

Drugs are potentially health-giving devices, which should not be abused such as by excessive or improper consumption, which can not only render them inefficacious but also damage health.

Finally, remember that a healthy life style is always infinitely better than any pill, because not even the best drug is capable of substituting for a fresh, varied and wholesome diet and reasonable and regular physical activity. A pill cannot repair the damage inflicted by harmful habits such as alcohol abuse, smoking, over-eating or excessive exposure to the sun or other substances which can harm our organism.

If You Want to Know More ...

Books

Atkins P., *What is Chemistry?*, Oxford University Press (2013).

Gerald M., *The Drug Book*, Sterling Milestones (2013).

Goldacre B., *Bad Pharma: How Medicine is Broken, and How We Can Fix It*, Fourth Estate (2012).

Mukherjee S., *The Emperor of All Maladies*, Scribner (2010).

Weinberg R.E., *The Biology of Cancer*, 2nd ed., Garland Science (2013).

Websites and Articles

Harris G., *Medicines Made in India Set Off Safety Worries, The New York Times* (2014), available at http://nyti.ms/1duVUbc.

Hsieh-Wilson L.C., Griffin M.E., *Improving Biologicdrugs via Total Chemical Synthesis, Science* (2013), available at sciencemag.org/content/342/6164/1332.short.

Janet Rowley, National Library of Medicine, available at nlm.nih.gov/changingthefaceofmedicine/physicians/biography_282.html.

John Vane, The Nobel Foundation, available at nobelprize.org/nobel_prizes/medicine/laureates/1982/vane-bio.html.

Leaf C., *Do Clinical Trials Work? The New York Times* (2013), available at http://nyti.ms/1DXhvH5.

Mintzes B., *Disease Mongering in Drug Promotion: Do Governments Have a Regulatory Role?, PlOS Medicine* (2006), available at plosmedicine.org/article/info%3Adoi%2F10.1371%Fjournal.pmed.0030198.

Mukherjee S., *Post-Prozac Nation: The Science and History of Treating Depression, The New York Times* (2012), available at http://nyti.ms/1AXPCvX.

Myers P.Z., *Chilling*, available at freethoughtblogs.com/pharyngula/2013/11/20/chilling/.

Olson H. *et al.*, *Concordance of the Toxicity of Pharmaceuticals in Humans and in Animals, Regulatory Toxicology and Pharmacology* (2000), available at sciencedirect.com/science/article/pii/S0273230000913990.

Paulson T., *Drug Development: Searching for Patterns, Nature* (2014), available at http://www.nature.com/nature/journal/v507/n7490_supp/full/507S10a.html.

Racaniello V., *HIV Gets the Zinc Finger, Virology Blog* (2014), available at virology.ws/2014/03/19/hiv-gets-the-zinc-finger/.

Rubin R.P., *A Brief History of Great Discoveries in Pharmacology, Pharmacological Reviews* (2007), available at http://pharmrev.aspetjournals.org/content/59/4/289.

Sir Alexander Fleming, The Nobel Foundation, available at nobelprize.org/nobel_prizes/medicine/laureates/1945/fleming-facts.html.

Sutherland W.J. Spiegelhalter D., Burgman M., *Policy: Twenty Tips for Interpreting Scientific Claims, Nature* (2013), available at nature.com/news/policy-twenty-tips-for-interpreting-scientific-claims-1.14183.

All Trials Initiative alltrials.net/.

The Berlin patient, available at en.wikipedia.org/wiki/The_Berlin_Patient.

Thomas K., *Why the Bad Rap on Generic Drugs?, The New York Times* (2013), available at http://nyti.ms/1OTXhj7.

Thomas K., *J.&J. to Pay 2.2 Billion in Risperidal Settlement, The New York Times* (2013), available at http://nyti.ms/1E4cyfz.

Myths to be Dispelled

1. A natural substance is less toxic than a chemically synthesised one

Whether a compound irrespective of its origin, natural or synthetic, is safe or harmful depends on how it interacts with the living organism after administration at a certain dose. All compounds are chemicals containing atoms. Without chemistry there would be neither synthetically derived compounds nor biologicals, but there would also not be any water, food or even ourselves. Among the most toxic substances around is arsenic, a chemical element which is 100% natural.

2. Each pill is a proper drug

This is not the case. Very many pills and health-related products possess properties of medicines but are definitely not drugs, some may even be harmful. Products such as vitamins, tonics, nutritional supplements, antioxidants, slimming aids, herbal extracts Bach flower remedies, ayurvedic medicines and homoeopathic remedies arrive in the pharmacy without having undergone laboratory tests, clinical trials or post-marketing control procedures after introduction into the market. They may contain harmful excipients and other substances which differ in quality or amount from those indicated on the information leaflet or label. Such pills may also differ in composition from one batch to another, given that the law does not require that they undergo quality control procedures.

3. Homoeopathic remedies may be less efficacious than traditional drugs but at least they do not cause harm

It is impossible to say which type of effect a medicinal product of whatever type including homoeopathic remedies may exert without appropriate tests and trials having been conducted. Homoeopathic remedies have not been experimented upon, as they mostly contain active agent at an extreme dilution, so that there is practically nothing but excipient. They arrive in the pharmacy without any validation of their presumed efficacy, safety or content. If a sick person in need of treatment takes a homoeopathic remedy instead of an efficacious drug which might resolve his or her health problems, he or she may aggravate the problem, and even death may occur because of inappropriate treatment.

4. Drug experimentation using animals is a waste of time

Studies in animals are indispensable to prove the safety and efficacy of a new drug. Some drug effects come to light only in a complete organism consisting of all organs which are reached by the drug and where it may be metabolised. These effects just cannot be observed in isolated cells used in experiments *in vitro*. We human beings are in many respects very different from animals, but we share with some of them, to a great extent, our evolutionary path and thus share with them many biomolecules which have been conserved to this day. Because of this molecular relationship, results from studies using animals which are sufficiently similar to us can suggest whether a drug may be tolerated by humans. Animal experiments are required by law before clinical trials can commence, and the law reflects the requirement demanded by consumers and patients that drugs which are to be prescribed by a medical doctor and approved by regulatory drug authorities are safe. Animal experiments are strictly regulated, safeguarding animal well-being and limiting their usage to the utmost minimum.

5. We know the mechanisms of action of all drugs

What we know about approved drugs (those validated for a therapeutic use by a drug regulatory authority) is to what extent they are safe and efficacious, but we know only for a few of them how they work.

6. Drug effects are the same in all persons who take them

The same drug administered at the same dose to different people generates blood levels which can vary up to 10–20-fold. No person is the same as another, and each drug can generate beneficial or detrimental effects in any and each individual. Efficacy and toxicity of a drug which have been studied in a limited number of people constitute median values, which may not necessarily be precisely reproduced in every individual.

7. The human body is like a machine and we know exactly how it functions

Machines come with instruction manuals because they were designed and put together bit-by-bit by human beings. In contrast, our body developed gradually by "trial and error" within about four billion years of the evolution of life on earth. Thus, the about 30,000 billion cells of which we are made up have been generated without a predetermined project plan. That is why the functions of the human body are known only in incomplete and imperfect fashion.

8. Molecular targeted drugs are better than traditional drugs

All drugs, from the oldest to the newest, have at least one molecular target, mostly more than one, otherwise they would not work. The fact that we do, in many cases, not know the molecular targets of drugs does not imply that these drugs are inferior to those the target for which we know.

9. Biological drugs are better than chemically synthesised ones

They are not better but fundamentally different. A simple cell into which our genes are inserted can generate molecules which are more sophisticated and refined than those generated by synthesis in the chemical laboratory. Compounds which are synthetically produced in a top modern chemistry laboratory are infinitely more simple and rough. However, biological processes cannot be controlled in every detail, because they arise from the tortuous and random path of evolution rather than from a detailed project designed by humans. The structures of compounds synthesised in the chemical laboratory are tightly controlled down to the last atom, because they are produced according to an instruction manual written by humans. Therefore, even though chemically synthesised substances may be less sophisticated than those produced by nature, we know almost everything about the former. We know how they were made in each detail and that they are absolutely pure.

10. Poisons and drugs are different from each other

The Swiss alchemist Paracelsus wrote in 1538: "All things are poisons, and nothing is without poison; only the dose makes a thing not a poison". His famous saying is still valid today. Even an innocuous substance like water can be toxic if one imbibes several litres of it within about an hour. *Vice versa*, potent poisons like alkaloids which occur in some plants can exert therapeutic effects when used at the right dose.

11. A new drug against a particular disease is always better than an old one

Each year, hundreds of new drugs are being approved. But that does not mean that they are really novel or better than those already in use. Sometimes, new drug molecules are only slightly modified versions of old drugs, the patent for which is about to expire, so some drugs are only "not inferior" with respect to the standard therapy in use for a certain disease.

12. Results of studies of new drugs constitute absolute certainties

The response to a drug varies considerably between individuals. It is possible that symptoms resolve regardless of any drug taken, given that many illnesses are self-limiting. It is possible that there is a placebo effect or that one of the many factors which can confound results of clinical studies plays a role in the response to a drug. Therefore, it is very difficult to decide whether a pharmacological effect can be attributed to the drug and not to confounding factors. The results of any experiment with a new drug, even the most rigorously conducted one, need to be seen as indications of general probabilities and never as certainties for an individual patient.

13. Drug safety is always advantageous for the patient

Safety is the primary evaluation criterion for the potential approval of a new drug. If a drug is found to cause harm, it is rejected out of hand even before one knows whether it is active or not. The priority allocated to safety is the response to the demands of consumers and patients who do not want to be exposed to risks. But is a high safety threshold always an advantage for patients? Or may the emphasis on safety entail the loss of the opportunity to approve a drug with efficacy against a disease against which we do not have any treatment yet, in spite of severe side effects of the drug? The question is an ethical rather than scientific one, and the right answer — if there is one — depends on societal and political attitudes with regards to health.

14. Generic drugs are worse than branded ones

Generic drugs are as good as their branded counterparts. By law, they must contain the same dose of the same active principle, and they function in the same way. But they cost at least 20% less than branded products. Generics are produced when the patent of branded drugs have expired. From this moment on, any company can produce and commercialise the generic form of a drug, the intellectual property of which is no longer protected. In most European countries, pharmacists are obliged to dispense the generic version of a drug available at the lowest price, a choice which benefits national health systems and saves societies large amounts of money. In order to improve the trust of the general public in generic drugs, their unfortunate name will shortly be changed to "equivalent" drugs.

15. Identification of a gene facilitates the understanding of its role in a disease and engenders the discovery of a new drug

The history of drug research provides ample examples of potential molecular targets, which have unfortunately not succeeded in generating new drugs. For example, cystic fibrosis is a genetic disease for which we have known for more than 25 years the mutated genes and the proteins causing the altered secretion in the lungs and other organs associated with the condition. Irrespective of optimistic expectations, efforts conducted for a quarter of a century have not furnished yet a satisfactory therapy. Many other diseases share this fate in that the abundance of molecular knowledge pertaining to these diseases contrasts with a desert of prospective useful treatments. The road to new drugs is really arduous.

16. A tablet is always better than an injection

Each mode of drug administration and each drug formulation have advantages and disadvantages. The choice depends on the health problem to be treated and on the characteristics of the drug to be administered. Tablets are convenient but have to transit acid in the stomach and alkaline in the intestine. In order to be able to pass the opposing charges of the two environments unharmed, tablets need to be neutral, that means lack any electrical charge. For drugs which require administration via injection, the active principle needs to be soluble in water or in a solution with the characteristics of body fluids. The formulation of a drug needs to ensure that the active principle is stable and reaches the correct part of the body at the right concentration and the right rate of delivery.

Did You Know That ...

Many drugs treat, a few cure, almost none repairs

There are about 14,000 different packages of medicines in the typical pharmacy, but only a small portion of these can eliminate the cause of a disease. The major portion deals with symptoms.

Insulin is generated by a genetically modified organism

Insulin, which today is taken by millions of diabetic patients, is produced in a biological microfactory, that is a bacterium or a yeast into which the gene for human insulin has been inserted. If you had diabetes, would you refuse to take recombinant insulin because it is the product of a genetically modified organism? Those who oppose any genetically modified organism (GMO) food should think about this.

The ancient legacy of the willow tree

Acetylsalicylic acid or aspirin has been used against fever, pain or inflammation since 1897, when the German chemist Felix Hoffmann working at the chemical company Bayer developed a better tolerated version of the active but somewhat harmful salicylic acid contained in the extract of the bark of the white willow tree. Salicylic acid and its modern version, acetylsalicylic acid, are among the very few ancient remedies — already Egyptians and Sumerians used the former one — which have undergone intense modern scientific investigation, so that we now know its mode of action.

"Anything but chemistry"

This was the reply of John Vane when interrogated on what he intended to do after he graduated with a degree in chemistry. He went on to win the Nobel Prize for his discovery of the mechanism of action of aspirin. Even the best researchers are sometimes disheartened, what is important is to pick oneself up.

Baby aspirin

Around the year 1982, two 30-year-old pharmacologists, Carlo Patrono at the Catholic University in Rome and Garret FitzGerald at Vanderbilt University in Nashville, Tennessee, were able to establish simultaneously in their laboratories that low dose aspirin (75–100 mg) can prevent the formation of thrombi. Today baby aspirin or aspirin cardio is prescribed to millions of people everywhere. The discovery has also changed the way in which cardiovascular diseases like infarction and stroke are treated.

The effects of viagra were discovered by accident

Viagra was investigated in patients with heart problems because it had been found to influence the vascular smooth musculature and blood flow. The idea was that it might inhibit the closure of the coronary arteries and thus prevent cardiac infarction. The study did not furnish the expected results as to heart disease, but some male patients noticed that the treatment provoked an unexpected collateral effect, that is a tendency towards penile erection.

Erythropoietin (EPO) is perhaps the best known biological drug

EPO is a hormone which is produced by our body to stimulate the growth of red blood cells. It is a small protein made up of 166 amino acids, the basic structure of which contains four branched sugar chains with the amount and composition of the sugars in each chain being variable and unpredictable. Therefore vials of EPO sold in pharmacies contain a heterogeneous mixture of molecules. Each has the same 166 amino acids, but the sugars differ. And the sugars here are more than the icing on a cake. The stability and efficacy of EPO changes according to the sugars present, even though the individual contribution of each sugar to the biological effects of EPO is unknown. EPO can generate red blood cells in a professional racing cyclist to make him or her go faster along steep roads, or it may be used in the treatment of a patient who has lost his or her red blood cells, because of the unwanted effects of cancer chemotherapy. The former use is illicit and dangerous because it is uncontrolled, the latter is permissible and potentially life-saving because it is administered under strict medical control.

In nature, poisons are the norm rather than the exception

Toxins offer a considerable evolutionary advantage to species of plants and mushrooms which are stuck to the soil and cannot escape an attack by predators. Poisons are common also in some marine animals which live fixed to rocks. When they feel they are approached by a predator, they emit their toxins into the surrounding water. The poisons found in the aquatic world are in general even more potent than those found in land-bound plants or mushrooms, because they need to exert their effect even after considerable dilution in water. It is not by chance that one of the most potent poisons is tetrodotoxin, produced by pufferfish. An antidote against this type of deadly toxin exists only in James Bond movies. Sometimes the most potent poisons if used in small doses are useful drugs and not harmful.

The first clinical trial

In 1747, James Lind, a doctor in the Royal Navy of the British Empire serving on HMS Salisbury, helped to deal with the umpteenth outbreak of the scurvy epidemic, an illness which seemed inevitable among crew members of any sea-bound vessel several weeks after embarkation. Dr Lind tried an experiment. He chose

12 of the mariners who had succumbed to scurvy and divided them into six pairs. Each mariner involved with the experiment had to stick to a predetermined diet, which was the same for all, and he also took a supplement which differed between pairs. The six supplements chosen by Lind were the agents which common belief at the time considered potential remedies against scurvy: lemon, orange, salt water, vinegar and a mixture of garlic, mustard and horseradish. After six days of treatment only four of the mariners, those who ingested lemon or orange, had improved, the others continued to deteriorate. This event is thought to be the first clinical experiment in history. We know today, that the cause of scurvy is the lack of ascorbic acid, or vitamin C, which occurs abundantly in citrus and other fruits and in fresh vegetables.

Participants in clinical trials are always volunteers

This regulation is part of the Nuremberg Code, a set of ethical principles established in 1947, after the Nuremberg trial of major perpetrators of Nazi crimes had brought the dismal cruelty and abuse to light committed against inmates of concentration camps. These unfortunate individuals were forced to take part in experiments which were not only inhumane, but also lacked any scientific value and rigour. Since then, nobody has been obliged to participate in a clinical trial against his or her will.

The placebo effect

Many patients react in a positive manner to a therapy irrespective of its components, because of the fact that they are preoccupied with their health problem. In these individuals, treatment with a placebo can elicit a perceived or actual improvement in their medical condition. This is a real and measurable effect which can be demonstrated in the laboratory or hospital and has been ascribed to the action of known molecules and areas in the brain. It can even occur in the absence of a therapeutic substance: sometimes words suffice to induce a placebo effect. Often pain or disease resolve themselves unassisted, and so it can be difficult to distinguish this natural development from a placebo effect. Comparing a drug with a placebo in a clinical trial can generate a clear result and show the true activity of the tested drug. If the activity of the drug in the study subjects is equal to that in those who take the placebo, it means that the effect of the drug is not different from that of an inert substance. It is however not always correct or ethical to compare the activity of a drug only with a placebo. If, for example, a drug already exists for the disease to be treated, ultimately the difference in efficacy is measured with respect to the existing treatment, and it would be unethical to offer only a placebo to patients in the control arm of the trial.

The nocebo effect

The nocebo effect is in a way the negative mirror image of the placebo effect. It is a detrimental physiological effect after drug administration which is not caused by the drug itself but by the patient's imagination concerning the effects or side effects of a treatment. Sometimes, very detailed information on the potential side effects of a compound can stimulate this effect, causing the occurrence of symptoms of which the patient has read or heard about. It is perhaps a good idea not to dwell too much reading the list of side effects on the patient information leaflet of a drug!

The reality about some novel anticancer drugs

The efficacy of a considerable number of novel anticancer drugs is rather limited when compared to drugs already in use for the particular disease. Some new drugs can improve survival time only by not more than a week. This fact contrasts starkly with both the claims sometimes made for the activity of a novel drug in advertisements and the staggering costs of many of this new type of drug. In reality, many novel anticancer drugs exert their efficacy only in a minority of patients who have a particular molecular characteristic rendering their disease amenable to the beneficial effect of the drug. In contrast, these drugs do not work in all the other patients with the same disease who lack this particular molecular characteristic. Survival data represent therefore a sort of middle way between two patient scenarios, increased survival for the few who have the right molecular trait, but survival equal to that associated with placebo or traditional drugs for the majority who lack the trait.

Recruitment of volunteers is difficult

According to a study conducted at the Fred Hutchinson Cancer Research Centre in Seattle, only 3% of patients with the appropriate characteristics accept recruitment into a clinical study, whilst 40% of such trials fail at the outset because of the inability of researchers to recruit enough suitable patients.

Many new drug dossiers are more voluminous than an encyclopaedia

In 2013, a large American Pharmaceutical Company presented to the Food and Drug Administration (FDA) the request for approval of a new sleep-inducing pill. The FDA is the US authority which clears the way towards the use and commercialisation of new drugs. The request submitted in electronic form consisted of 41 gigabytes, compared to the 9.2 gigabytes which all Wikipedia entries in 2013 took up. This dossier for the sleep inducer is not an exception. Efficacy and safety of each new drug which seeks approval have to be described in such detail that the documentation generated can exceed encyclopaedic proportions.

New drug approval in Europe

The European Medicines Agency (EMA) currently situated in London is the only institution which can authorise the use of a new drug in the countries of the European Union (EU) after a process of evaluation which should not take longer than 270 days. Drugs approved by the EMA can by right be sold in all EU countries, and no country can oppose this decision. After approval by the EMA, the appropriate national drug regulatory body of each member state has the task to establish if the drug merits inclusion into the national formulary and reimbursement by the country's national health service.

Thalidomide: disaster and rebirth

Introduced into the market in 1956, thalidomide was approved for medical usage in Great Britain in 1958. By 1961, it was used in at least 100 European and African countries as antisickness drug in pregnant women. In 1960, the request for approval to enter the American market led to a study in which 2.5 million tablets were distributed to about 20,000 patients. In the course of this trial, there were indications of grave deformations in 17 newborn babies. So authorisation of the drug was rejected in the US. In the countries in which thalidomide had been widely used, more than 10,000 children were born with phocomelia, i.e. deformed arms or legs, before the drug was eventually withdrawn from the market in 1961. Italy managed to avoid this tragedy as its national drug regulatory body had been late initiating the approval process. The thalidomide disaster focused the attention of regulatory bodies on teratogenesis, the capacity of drugs to alter the development of the embryo. From the 1960s onwards it was stipulated by law that the potential teratogenic effects of any new drug need to be evaluated in studies in those animal species, the gestation period of which resembles that of humans. In the 1990s thalidomide experienced an unexpected revival, because of its beneficial effects on multiple myeloma. This is an example of the fact that many poisons are also remedies, depending on the dose, the disease and the patient who is to receive the agent.

To what extent are results of clinical studies made public?

In the US the FDA can permit access to the documentation on a new drug after it has evaluated the motivation which led to the request for information. In contrast, in Europe drug dossiers are secret documents. The initiative, AllTrials instigated by the English epidemiologist Ben Goldacre proposes obligatory publication of the results of all formally authorised clinical trials including those which yielded negative results.

Do drugs against depression act like a placebo?

This seems to be the case for some of the most frequently used antidepressants, the selective serotonin reuptake inhibitors (SSRIs). Prozac is the commercial name of the best known of these drugs, with its active principle being fluoxetine. Prozac has been on the market for more than 25 years. This type of drug is active in only 20–25% of patients who suffer from the most severe form of depression. But in the majority of the patients with other depressive conditions they do not work any better than a placebo. We are talking here about millions of people. In the Western world, about one in 10 citizens receive prescriptions of fluoxetine or related antidepressants. The companies which marketed Prozac and its pharmacological relatives knew from the outset about the limited efficacy of these drugs, but kept trial results secret, as is permitted by law.

Individual preferences of medicines

In Japan, few would take a tablet coloured dark grey, green or pink. In other countries, red black or colours recalling sweets are not popular. Tablets are not sold in single units but in packages and also in this respect countries differ in preferences. Americans prefer bottles containing about thirty tablets, the rest of the world is now used to blister packs made of plastic and aluminium.

Only one in 60,000 experimental drug molecules goes on to obtain regulatory approval

Each year more than a million and a half experimental drugs undergo preclinical studies in the laboratory. Of these, about 300 go on to be subjected to clinical trials, and fewer than 25 turn out to be promising and receive approval. The time between the start of clinical experimentation and approval tends to be about eight years.

Pharmacovigilance can improve therapy

The rotavirus causes a form of diarrhoea which kills more than 400,000 children every year, especially in developing countries. About two million children are brought to hospitals per year. The first vaccine against rotavirus was approved in 1998, and many children were immediately vaccinated. Within a few months a rare intestinal problem emerged in about one in 12,000 vaccinated children, and this effect had not been observed in the earlier extensive clinical evaluation of the vaccine. After pharmacovigilance had indicated a problem, the vaccine was immediately withdrawn, and in 2006, a new version lacking this side effect was approved and distributed. In the development of the new version, researchers had to start right from the beginning. The vaccine had to be tested in a population of non-vaccinated children in whom safety was confirmed. Approval of each drug is valid for a specific compound to be used against a specific problem in a well-prescribed population of patients. As soon as any of these parameters change, one has to go back to the experimental starting point.

Any drug targeted against a microorganism is destined to fail

The capacity of a population of viruses or bacteria to change their genetic armamentarium is considerable, and thus they tend to acquire resistance against their target in at least some individuals on therapy. It is much more difficult for microorganisms to develop resistance against several drugs administered together. Today, the number of people who die each year of acquired immunodeficiency syndrome (AIDS) is declining, and 34 million survive infection with the virus thanks to the availability of a combination therapy called *highly active antiretroviral therapy* (HAART). This consists of one single pill containing several agents active against human immunodeficiency virus (HIV). HAART has decreased the rate of deaths from AIDS by 50–70% and has transformed this disease from a death sentence to a chronic condition.

Many drugs are prescribed for off-label indications, for purposes not listed in the information leaflet

The approved indications of a drug are the specific health problems against which the drug has been evaluated and against which its use has been approved. Irrespective of the approved indications of a drug, a medical practitioner can prescribe the drug for any medical purpose, as long as he or she considers the use as safe and efficacious on the basis of his or her knowledge and experience. In the US, one drug in five is prescribed for off-label use. However, a pharmaceutical company is not allowed to advertise off label use to medical doctors.

The medical world pre-anaesthesia was brutal

Only two hundred years ago, patients arrived on the surgeon's operating table inebriated with alcohol and whatever substance was able to knock them profoundly out so that they could be operated upon under conditions of immeasurable pain and amid awful amounts of blood. Surgery was restricted to rare occasions and patients with little hope of recovery. Extreme speed of the operation was essential, so that an amputation would have been completed in less than a minute, this being the only way to avoid the patient dying of excruciating pain. Before the advent of anaesthetics, the pitiful cries of patients undergoing surgery filled operating theatres, then followed by ominous silence. Today, cries of pain are heard only in delivery wards in hospitals, which sometimes still discourage the use of anaesthetics for women in childbirth. Without anaesthetics, organs like the abdomen, thorax, joints and of course heart and brain were out of reach of surgical practice, which at the time dealt only with the most external parts of the body and left the rest to be dealt with by so-called internal medicine. Routine operations, performed daily in hundreds of thousands of patients, like coronary bypass surgery or hip replacement, would have been be unimaginable. Anaesthetics were used for the first time in the dental practice of William Morton in Boston in the mid 19th century. They were then distributed to operating theatres everywhere, where they truly changed the condition of humanity.

A healthy life style is always better than any pill

Not even the best drug is capable of substituting for a healthy, fresh and varied diet and regular physical activity. Drugs can never repair the damage inflicted on the body by detrimental habits such as too much alcohol, cigarette smoke, overeating, too much sun exposure and ingestion of other substances which can damage our organism.

Index